A Critical Introduction to Properties

A Critical Introduction to Properties

SOPHIE R. ALLEN

Bloomsbury Academic
An imprint of Bloomsbury Publishing Plc

B L O O M S B U R Y
LONDON · OXFORD · NEW YORK · NEW DELHI · SYDNEY

Bloomsbury Academic

An imprint of Bloomsbury Publishing Plc

50 Bedford Square	1385 Broadway
London	New York
WC1B 3DP	NY 10018
UK	USA

www.bloomsbury.com

BLOOMSBURY and the Diana logo are trademarks of Bloomsbury Publishing Plc

First published 2016

British Library Cataloguing-in-Publication Data

A catalogue record for this book is available from the British Library.

ISBN:	HB:	978-1-4725-7560-9
	PB:	978-1-4725-7559-3
	ePDF:	978-1-4725-7557-9
	ePub:	978-1-4725-7558-6

Library of Congress Cataloging-in-Publication Data

Allen, Sophie (Philosophy Lecturer)
A critical introduction to properties / Sophie Allen.
pages cm.– (Bloomsbury critical introductions to contemporary metaphysics)
Includes bibliographical references and index.
ISBN 978-1-4725-7560-9 (alk. paper)– ISBN 978-1-4725-7559-3 (alk. paper)–
ISBN 978-1-4725-7558-6 (alk. paper) 1. Whole and parts (Philosophy) 2. Universals
(Philosophy) 3. Tropes (Philosophy) I. Title.
BD396.A45 2016
111'.8–dc23
2015028328

Series: Bloomsbury Critical Introductions to Contemporary Metaphysics

Typeset by Fakenham Prepress Solutions, Fakenham, Norfolk NR21 8NN
Printed and bound in India

Contents

Acknowledgements vi

1 Introduction 1

2 Universals 7

3 Tropes 39

4 Properties as sets or resemblance classes 67

5 Properties: Grounded or ungrounded? Sparse or abundant? 93

6 Intrinsic and extrinsic properties 113

7 Properties and their causal role: Categorical and dispositional properties 139

8 Causes, laws and modality 167

9 The ontological status of properties 191

10 Conclusion 217

Glossary 219
Bibliography 221
Index 231

Acknowledgements

While it would be commonplace at this point to say that this book would not have been possible without the help of certain people, that statement would be false. What is true though is that it would not have been actual in its present form without their generous help. So, thank you to: Alastair Wilder for extensive comments and proofreading; Ben Page; Leonie Smith; Anna-Sofia Maurin for inviting me to discuss properties in Gothenburg; and Karen Bennett, Dan Giberson, Barbara Vetter, Javier Cumpa and Anjan Chakravartty for giving me early access to forthcoming work. Thank you also to Chris, Freya and Connor.

1

Introduction

This chapter introduces the main aims and themes of the book: the ontology of properties and the explanatory work which properties can do. The general problem of qualitative sameness is explained, and properties are compared with relations. Some central influences upon the formulation of a theory of properties are introduced, including the commitment to abstract objects, the account of modality in play, and the relationship between epistemology and metaphysics.

1.1 What?

What do blue things have in common? Or electrons? Or aardvarks? Or planets? Or prime numbers? What determines whether a particular belongs to one kind rather than another? We make judgements about resemblance in specific cases, and make yet more comparisons between seemingly very different kinds of things: both aardvarks and planets have mass, shape, and location, while electrons have arguably only the former of those three and prime numbers have none. Underlying our judgements of qualitative similarity and difference is the general thought that there is something which makes distinct particulars qualitatively the same as, or similar to, each other and since Plato, it has been a respectable philosophical pastime to try and work out what these things are. Let us call them *properties*, although that tells us very little as it stands.

In addition to properties, we can also classify pairs of particulars according to the relations they instantiate: Oxford is to the west of London, and so is Cardiff; the pairs <Oxford, London> and <Cardiff, London> are similar to each other in a particular respect; and they share that with <Akron, New

York>, <Ahmedebad, Kolkata> and indefinitely many more ordered pairs of locations on Earth. There are obviously many aspects of the cities themselves which make them differ too, properties which one city has and another lacks. But the relation of *being west of* is a quality instantiated by the pairs of cities regardless of this. Relations are not restricted to being binary relations which hold between pairs: they can hold between any number of objects. In families of three siblings, someone has the property of *being the middle child*; that is, someone was born between the other two children, giving a three-place relation. Stuart *is born between* Mark *and* James and Nick *is born between* Jane *and* Julian, forming the ordered triples <Stuart, Mark, James> and <Nick, Jane, Julian>. But a formalized version of the English relation of *being born between* might also turn out to be five place, or seven place, or nine place, and so on, depending upon the number of siblings there are in a family.

There are some specific questions we can ask about relations: one might wonder whether the three-place relation is different from the nine-place one in examples such as the one above, or whether we should countenance multigrade relations which can hold between varying numbers of particulars; one might want to know whether a binary relation has certain properties, such as being transitive, symmetric, or reflexive; or whether a relation holds in virtue of the intrinsic nature of the particulars it connects, or if it is an external addition to the ontology. But in the main, the metaphysical concerns we have about relations are the same as those which we have about properties. (Astute readers may notice that yet more properties or relations – this time properties or relations of properties and relations – will underlie this observation too.) So, in view of their similarities, I will discuss properties and relations together, mainly talking in terms of properties unless the difference is important.

1.2 How?

The aim in the first half of the book is to explore a range of ontological options about what properties might be: universals, tropes, resemblance classes and natural classes. If properties can be grounded in another kind of entity, the answer to the question of what that entity is, or those entities are, will be found there. Within these four broad categories are more specific and nuanced views, some of which will be cast by the wayside for logical reasons, or mended to avoid philosophical objections until they are able to stand up to scrutiny as well as the other theories on offer. The ability of metaphysicians to adapt their theories in the face of criticism will become apparent: theories which begin as weak, intuitive glosses on what properties might be develop into sophisticated accounts of qualitative resemblance between particulars and whatever that involves. But it is precisely this adaptability which makes it difficult to choose between the theories, in part because of the apparent tendency of theories to evolve in order to minimize the differences between them. Each theory of properties sets out to answer a very similar array of questions and to solve similar problems, and so it is no surprise that the surviving theories are equivalently useful for explaining what properties are and providing entities which do what properties are supposed to do. After all, they have been designed that way.

Furthermore, some of the most interesting philosophical differences between accounts are interesting precisely because they involve disagreements about very different areas of philosophy, only tangentially related to qualitative similarity and difference. For instance, a major division is created by the acceptance, or the refusal to accept, the existence of abstract objects. A second, by adherence to the Principle of Instantiation; the view that the only properties which exist are those that are, have been, or will be instantiated, so that uninstantiated properties do not exist. The third division crosscuts the second and involves a commitment to actualism, restricting what exists to what actually exists. If one combines this with the Principle of Instantiation, then properties are further restricted to those instantiated in space-time. Fourth, the outcome of the debate about the ontological status of properties is influenced by much broader matters concerning what the relationship should be between metaphysics and epistemology, the relationship between what there is and what we can know. On the one hand, metaphysics is practised by those who care little for whether we can experience or otherwise discover the entities they claim to exist, while on the other, some philosophers have strong reservations about accepting the existence of entities when we have neither a clear idea about their nature nor a means of finding it out. For the most part, this book will go about its business in feigned ignorance of the latter viewpoint, treating metaphysical theorizing as being about an objective reality which would exist whether or not we were here to discover it. However, I will have more to say about the how we know about properties, and what their ontological status should be, in the final chapter of the book.

Having discussed different accounts of the ontological basis of properties, the latter half of the book will investigate some general metaphysical issues about the nature of properties, some important distinctions between different conceptions of properties, and the explanatory work which properties can do, such as how they are related to other phenomena including causation, laws and modality. Whether or not we are successful in giving an account of properties in terms of something else, there are some questions which apply more generally to properties whatever they are. For instance, one might think that we can learn something useful about the members of an ontological category by knowing what makes them numerically the same as each other, or different. In Chapter 5, I will investigate whether there are constitutive identity and individuation criteria for properties. This search leads to another concern, since the plausibility of an identity criterion for properties depends in part upon how many properties there are. Are there enough distinct properties for every type of thing which is logically possible? Or enough to provide linguistic meanings? Or is the ontology of properties sparser than that, providing just enough properties to determine the kinds of entities and the causal interactions of the actual world? To some extent, the answer to this question turns out to depend upon what properties do in our metaphysical theory: if we want enough properties to give an account of the meaning of predicates, we may opt for a more abundant ontology of properties than if we think properties are an essential feature of the causal interactions in the world and that all causal interactions can be brought about by a minimal fundamental set of properties. However, this alignment is not set in stone: one might think that there is an abundance of properties involved in causation; or that the abundance of meanings can be accounted for by their ultimately depending upon a minimal set of fundamental properties which determine or generate all the rest.

Chapter 6 is perhaps the most technical, being an investigation of recent attempts to draw a distinction between intrinsic and extrinsic properties. There are two principal strategies by which this can be done: first, using criteria which try to stick to logical concepts; and second, those which rely upon metaphysical notions such as grounding, or specific features of properties, and thus require ontological commitment in addition to logic. This chapter also introduces a range of ways of classifying properties themselves: those which are purely qualitative; those which require the existence of another entity to be instantiated; and properties which are associated with necessary entities, including those which are instantiated by everything which exists.

Chapter 7 examines the relationship between a property and its causal role: Is a property essentially what it can do, or can it change its causal role in different possible situations? The main debate here is, on the one hand, about the intelligibility and implications of the opposing, essentially qualitative or quiddistic view of properties and, on the other, about whether quiddistic properties are needed in order to characterize possibilities which cannot be articulated in terms of essentially causal properties. Once again, philosophical views about modality, and about what is and isn't possible, impinge upon the discussion about what the best formulation of a theory of properties is.

Chapter 8 explores the relationship between properties and other phenomena including causation, laws of nature and modality. It turns out that certain conceptions of properties – those which treat properties as essentially causal entities – have a greater capability to produce a more extensive and unified metaphysical theory which is parsimonious in comparison to its rivals. Such essentially causal, dispositional properties work in this way because it is possible to give an account of how both laws of nature and modality are defined in terms of them. Furthermore, it might be possible to extend this modal reach in order to cover logical possibilities as well. The cost of this may be a restriction on what is possible which some will find untenable, and in an echo of Chapter 7 this may prompt a move back to a quiddistic conception of properties.

The final chapter takes a look at the ontological status of properties and raises objections to the realist account of properties which has been presupposed throughout the book. Although it is plausible enough to assume the objective existence of properties due to the important metaphysical role which they serve, we face a sceptical challenge having postulated these entities about whether we can find out which properties there are. We can react to this observation with denial of the sceptical conclusion, or humility towards it, or else with rejection of mind-independent properties themselves. The proponents of the latter option owe at least a sketch of what the theory of properties might look like if properties are not objective aspects of a mind-independent world.

1.3 Why?

One thing that this book will not give you is a definitive, unequivocal answer to the question of what a property is. What it aims to do is to provide you with the tools to answer that question for yourself. As I have already noted, many of the apparent philosophical chasms between theories can be bridged once we realize that the theories do as well as each other in providing a workable philosophical account of properties in which properties do everything that we hope they will and perhaps more besides. The choice between such theories is dependent upon too many interconnected issues in metaphysics, which I have no opportunity to discuss in this volume, upon the relative weighting accorded to theoretical virtues – how one chooses to play off parsimony and simplicity against explanatory power, for instance – and upon decisions about which claims one is prepared to take as primitive judgements on the basis of pretheoretical intuitions. One might ultimately decide that if there are no rational grounds upon which to make a choice between ontological theories, then this implies that the theories collapse and that, despite the apparent plurality, there is only one theory after all; but that is not a philosophical position that I will argue for here. Rather, I will leave it to the reader to draw his or her own conclusions about what the best theory of properties is.

Readers should note that although this book is written to be suitable for someone with only a basic grounding in philosophy, many of the topics covered here are far from introductory and several are drawn from contemporary metaphysical debates which may make better sense when given time to settle. We might look back with the benefit of hindsight and realize that some problems were not worth worrying about, while others which were skipped over in this book deserved more space. Furthermore, it would have been easy to find the material to write an individual book to cover the subjects of each and all of the forthcoming chapters and so some influential and interesting objections, and whole debates, have been entirely missed out. Further reading is suggested for those who want to explore topics in more depth and also to point in the direction of some of the more complex material which was denied inclusion due to lack of space.

2

Universals

This chapter introduces the distinction between particular and universal and discusses theories which take exact qualitative similarity to consist in identity by maintaining that qualitatively identical particulars instantiate the same universal. Semantic and ontological arguments for the existence of universals are considered—including the 'one over many' argument—as well as Russell's resemblance regress which claims that alternative views of similarity must postulate a universal resemblance relation to avoid regress. The question of whether universals should be immanent or transcendent is addressed, including: Russell's argument that the immutability of universals precludes their being spatio-temporal; whether abstract entities are more mysterious than those which can be wholly present in distinct spatio-temporal locations; and whether transcendent universals are required to account for uninstantiated properties, alien properties, and inexact resemblance between particulars. I examine whether universals theorists can avoid a regress of instantiation similar to Russell's resemblance regress, and whether universals can instantiate themselves.

Is it literally true that qualitatively similar particulars have something in common? Supporters of the first ontological account of properties I will explore maintain that it is: when distinct entities share a property, they instantiate, or participate in, one and the same universal. For example, the same universal *blue* is instantiated by any random collection of distinct blue things – the cloudless daytime sky this morning, a patch of International Klein Blue paint, Elvis Presley's suede shoes, a particular forget-me-not, a mountain bluebird, and the broken umbrella discarded next to my desk – and this is what makes them all blue. Qualitative similarity is explained by identity: the blue of each particular is one and the same. Similarly, every individual entity with a mass of 1 kg instantiates the universal of *having mass of 1 kg*; each particular just act participates in the universal *justice*; and every individual dog instantiates the universal *dog*.

Whereas particular entities are unrepeatable – an individual dog, or a particular 1 kg bag of sugar, can only occupy one region of space-time each – universals are *repeatable*, they can be present or instantiated at many different spatio-temporal regions; that is, they can be *multiply exemplified*. The existence of universals, it is claimed, explains how the same feature can be shared or exemplified by many different particulars in different times and places, and gives an account which fits with our intuitions about sameness of kind.

Two aspects of this rough-and-ready picture need some refinement to avoid misunderstandings from the start. First: one need not be committed to the claim that every single predicate has its own universal; there may not be a universal for every possible linguistic description, including ones for all, or even any, of those in the examples used in this chapter. In parallel with many accounts of the ontology of properties which we will come across in forthcoming chapters, one might think that some universals ground others – that is, that some universals are more fundamental than others – or that some predicates do not pick out universals at all. Thus, when we say that each of a group of things 'is blue', or (of other individuals) that each 'is a dog', we may be saying something far more complicated than 'there is a universal *blue* which the former all share', or 'there is a universal *dogginess* which the latter group share': blueness and dogginess need not be fundamental or 'genuine' universals (Armstrong 1978b), but might be grounded in other universals and the relations between them. For example, someone of a physicalist persuasion might maintain that only physical universals exist – perhaps only a subset of physical universals more fundamental than the rest – and that all other universals are determined somehow by these. Because similar issues arise in relation to other accounts of properties, questions about how abundant universals are will be postponed to Chapter 5.

The second issue concerns the different terms used for the relationship between particulars and universals, saying that individuals 'instantiate', or 'participate in' a universal. Moreover, in addition to a particular 'participating in' or 'instantiating' a universal, the particular might also be said 'to exemplify', 'to partake of', or 'to exhibit' the universal too.[1] Although there is disagreement about the nature of the relationship between particulars and universals (of which there is more discussion in 2.4), these different opinions do not cut along terminological lines. Let us stick with 'instantiation' for now, with 'exemplification' and 'participation' set aside for occasional use.

So far, we know that universals can be multiply instantiated – a single universal can be instantiated at different times and places – in contrast to particular entities, each of which occurs at only one spatio-temporal location. Furthermore, while a concrete particular excludes all other particulars of the same ontological category from its region of space-time, many universals can be instantiated within the same spatio-temporal region, that is, they can be simultaneously instantiated by the same particular. For instance, one particular dog can be *black*, *wet*, *smelly*, *have a mass of 80 kg*, *be diabetic*, *omnivorous*, *a Great Dane cross* and so on, with all these universals instantiated at the same time and place, but no other particular dog – nor any other particular thing for that matter – can coincide with exactly the same spatio-temporal region as the aforementioned dog.

A third metaphysical contrast between universals and particulars might be drawn in terms of instantiation: whereas particulars can only instantiate universals, universals can both be instantiated and instantiate other universals (and perhaps also themselves, see 2.5). My broken umbrella instantiates a particular shade of cobalt blue (and the particular umbrella cannot itself be instantiated), but that universal *cobalt blue* is not only instantiated, but it also instantiates the universals *blue* and *colour*. Although this third contrast is sometimes used to mark the distinction between particulars and universals, I will not rely upon it (Lowe 2002a: 350–2). One drawback with this criterion is that it restricts which universals exist to those which *could be instantiated*. While this might be considered plausible, and is not as restrictive as saying that the only universals which exist are those which are instantiated at some point in space and time, one might want to allow the existence of universals such as *round square*, or *non-self-identical*. At this stage, I do not want to rule these out.

But why should we believe in the existence of universals? I will consider three kinds of argument for their existence.

2.1 Why believe in universals? Arguments for their existence

2.1.1 Metaphysical and semantic arguments

The first argument for the existence of universals can be traced back to the work of Plato and is known as the 'one-over-many argument' (although that is not a term which Plato used). Plato called universals 'forms' or 'ideas' (although he did not mean 'idea' in any psychological or subjective sense), and he postulates their existence in order to give an account of how predicates and general terms get their meaning: 'We customarily hypothesize a single form in connection with each collection of many things to which we apply the same name.'[2] For example, many diverse particulars can be called 'beautiful': a particular statue, a mountain, a person, a poem, entities which are so different from each other they do not readily reveal a common basis for their beauty. However, the universals theorist thinks that such a common feature must exist. What makes these different things beautiful – and the term 'beautiful' apply to them – is their participation, or sharing in, the universal, or form of, *beauty* itself. Similarly, but more prosaically, particular tables instantiate the universal *table*, individual people instantiate the universal *person* and some of them also instantiate the universal *red-headed* and some the universal *left-handed*.

Here we can distinguish a semantic or linguistic aspect from an ontological aspect of the one-over-many argument: universals help to provide an account of the reference of predicates and general terms such as 'is blue' and 'table', and thus explain how such terms get their meaning and contribute to the truth-value of sentences (that's the semantic bit); and universals also give an ontological account of what it is that makes distinct particulars qualitatively similar or the same. The two aspects are not unconnected because, according to this account, what there is – namely which particulars there are instantiating which universals – determines which sentences are true. However, some philosophers emphasize the ontological aspect: one might be interested in what it is that grounds similarity between distinct particulars and what fixes their qualitative nature, whether or not we can refer to such entities in our ordinary natural language. The ontological aspect of the argument can be considered independently without trying to draw any linguistic conclusions. For others, the account of predication afforded by the existence of universals is of primary importance. These differences in motivation may ultimately lead to preferences for different accounts of universals later on.

The account of predication given by the semantic aspect of the one-over-many argument can be fleshed out in the following way. When we make a statement such as 'Socrates is wise', we are attributing a predicate '... is wise' to a subject 'Socrates'. Although there is plenty of disagreement in the philosophy of language about how proper names, or singular terms, are to be treated, it seems plausible to say that Socrates, the particular man, has a lot to do with what the name 'Socrates' is doing in the sentence: Socrates is the referent of 'Socrates' and sentences which are about Socrates have their truth-value and (perhaps) their meaning in virtue of the man himself. But what is '... is wise' doing in the sentence? It is not obvious, as it was in the case of the singular term 'Socrates', that predicates have referents, and thus where the meaning of '... is wise' comes from and how the whole sentence 'Socrates is wise' turns out to be true. The universals theorist suggests that the simplest solution is to treat predicates as referring expressions in a parallel way to singular terms: predicates refer to universals and can be treated like singular terms such that saying that 'Socrates is wise' refers to the universal *wisdom* which he instantiates; now, all the terms of the sentence have reference and we have an account of how the sentence means what it does in terms of the entities which make it true. Moreover, such an account of predicates in terms of universals permits us to account for their generality: if 'Socrates is wise', 'Plato is wise' and 'Anaximander is wise' are all true, the attribute of wisdom refers to the same thing in each case, not to an entity or quality which is individual to each of them.

Universals can also explain the meaning of general terms, such as 'courage', 'hot', 'blue', 'tree', 'wisdom' and so on. Some of these terms are problematic because they appear to be referring to rather abstract qualities, such as *wisdom* or *courage*, and we might find it difficult to agree about what counts a particular instance of wisdom, never mind the quality of wisdom in general. But, even when we consider more mundane general terms of which we can easily identify instances, such as 'tree', 'blue' and 'hot', they are notoriously difficult to deal with if one tries to give an account of meaning in terms of sense experience, since it appears that our sensory experiences are always of particular entities.[3] I cannot have an experience of trees-in-general, just of this particular tree or that particular tree, each of which has a specific size, shape, range of colours and so on, and since these qualities are different in different trees, none of these specific features is essential to something's being a tree. Thus, it is not clear how the meaning of our general term 'tree' is extrapolated from these individual cases, although some philosophers have argued that it can be. The universals theorist argues that we should accept universals as the referents of general terms: 'tree' refers to the universal, and the particular instances of trees are instances of trees because they instantiate this universal. Other, more empirically elusive terms such as 'courage' and 'wisdom' can be dealt with in the same way (although some philosophers might choose to say that they have no referents). Moreover, the account can be linked to that suggested for the reference of predicates above: the category of universals permits one to give a unified account of the semantics of predicates and general terms.

To be effective, the ontological and linguistic arguments for the existence of universals presented so far – all of which are loosely based on the one-over-many argument – need to be supplemented with negative arguments directed against alternative views: nothing which has been claimed above rules out competing accounts of the ontology of properties, or of the semantics of predication and general terms, which can do the jobs that universals have been postulated to do. Of course, one might still find the account of universals compelling despite this, although there are still many details of the account to be worked out. But only when the universals theorist has presented serious or insurmountable difficulties with the theories of his rivals would he have shown more than the initial plausibility of his favoured ontology as a basis for properties, and this negative project against alternative views is one of the key projects which some universals theorists have undertaken.[4]

2.1.2 *The resemblance regress*

One family of negative arguments seeks to establish the indispensability of universals: alternative accounts of properties – whatever these may be – require the existence of at least one universal and thus it turns out that every theory of properties is a universals theory after all. Versions of this argument are presented by Russell (1912a: 55) and Armstrong (1978a). For brevity, I will focus upon one competing account of properties to which this form of reasoning is applied.

Recall that properties and relations are to be treated alike. Russell admits that it is immensely difficult to show that universals are the entities picked out by adjectives such as 'red', 'triangular', or 'saline', or by substantives such as 'water', 'justice' and 'energy'; but, he argues, universal *relations* are required, even by those who would like to deny their existence. For example, consider particular triangles a, b, and c. The universals theorist claims that a, b and c each instantiate the universal *triangularity*, but one might deny this and say instead that triangles a, b and c *resemble* a certain triangle d (which we have picked as an exemplar), or resemble each other, in a certain way.

We can put the matter formally as follows. We want to answer the question of why a, b, and c are all members of the set of triangles. We try to account for individual cases of resemblance between the particular triangles a, b, and c, by saying that each particular is 'related' to each other particular, giving pairs <a, b>, <b, c>, <a, c>. (I will ignore the symmetry of resemblance for simplicity, except for the complication noted below.) But now we face the question of why <a, b>, <b, c> and <a, c> are all members of the set of resemblance relations. If that is because <a, b>, <b, c> and <a, c> resemble each other in the manner just described, then this account demands that this be explained by their falling into pairs: <<a, b>, <b, c>>, <<b, c>, <a, c>> and <<a, b>, <a, c>>. But why do these pairs resemble each other? If further pairs are required, of the form <<<a, b>, <b, c>>, <<b, c>, <a, c>>> and <<<b, c>, <a, c>>, <<a, b>, <a, c>>>, the regress in this account becomes apparent. Every pair of resembling particulars must be 'related' to the others by appearing in further pairs. Moreover, once we account for the symmetry of resemblance, it becomes clear that each stage of the regress involves twice as many pairs as the order below it contains.[5] But if we say that these higher-order cases of resemblance are just the same as before, then we have introduced a repeatable entity – a universal relation – which accounts for the similarity between the different pairs, the pairs of pairs, the pairs of pairs of pairs, and so on.

If the resemblance theorist argues that the similarity between individual instances of resemblance is itself to be explained in terms of resemblance between instances, she leads her theory into an infinite regress. To avoid the regress, resemblance must be treated as a universal relation. Those who attempt to do without universals and talk only in terms of resembling particulars have been forced to employ a universal after all and as Russell puts it 'we find that it is no longer worth while to invent difficult and unplausible [sic] theories to avoid the admission of such universals as whiteness and triangularity' (1912a: 55). It is simpler and more perspicuous to accept the category of universals from the start.

This argument makes a strong case for the existence of universals on the basis of their indispensability to a competing account of qualitative similarity. Moreover, the regress argument can be generalized to object to other ontological views. Those who do not favour a straightforward theory of universals will find themselves forced to postulate universals anyway to avoid regress, thereby undermining their own position.

But perhaps this argument is not as strong as it first appears. For a start, it begs the question: the universals theorist has presupposed that the *universal* relation resemblance is the only way in which resemblance can be characterized; he has assumed exactly what he is trying to prove. But, what of the regress? Surely that rules out alternative accounts? The answer to this is more complicated and we shall revisit the resemblance regress in 2.4, when there is a better basis upon which to judge how seriously the threat of regress should be taken.

2.1.3 *Universals at work: Laws of nature*

The third strategy used to support the existence of universals emphasizes their utility for specific metaphysical tasks, most frequently to give an ontological account of laws of nature. According to this view, laws of nature are not simply linguistic descriptions generalizing over individual cases of the form *Any particular that is F is G*, nor are they regularities holding between all the individual instances of F and G together which make such universal generalizations true; instead, laws of nature are necessary connections which hold between universals of the form *Necessarily (F,*

G), or *F-ness necessitates G-ness*, which entails that everything which instantiates F must also instantiate G. The strength of the 'must' here may fall short of metaphysical necessity, such that which laws there are may vary between possible worlds; laws conceived of as relations between universals may be governed by natural or nomological necessity, and yet be metaphysically contingent. This account is argued to be more plausible than those which characterize laws as universal generalizations or regularities, since the necessity involved allows us to distinguish law-governed correlations from accidental ones with which the regularity account has difficulty. Whether this characterization of laws as necessary connections between universals is correct is controversial however, and deserves more space for debate than it can be permitted here. Since it depends in part upon which conception of properties is preferable, there will be reason to return to it in 8.4.

2.2 What are universals? Immanence and transcendence

Let us suppose that we now have reason to think that universals exist. What species of entities are they? There are different conceptions of universals, with an important distinction being based upon whether universals are located in space and time (or, from here on, space-time): Are universals located where the particulars which instantiate them are located? Or are they abstract entities, which are not located in space-time?

An affirmative answer to the former question yields the account of *immanent universals* championed by David Armstrong: a universal is a repeatable entity, *wholly present* in each spatio-temporal location occupied by its instances. On the other hand, the claim that a universal is not located where its instances are located – in fact, it is not located anywhere – results in the doctrine of *transcendent* universals, universals which lack spatio-temporal features entirely. Is one conception of universals preferable to the other?

Before attempting to answer this question, it is worth noting that there is little agreement about what to call two distinct views about universals which I have just outlined. I will stick to calling them 'transcendent' and 'immanent', but other writers distinguish universals into 'Platonic' and 'Aristotelian' respectively (although the view may not have been Aristotle's own), or use the Medieval Latin '*universalia ante rem*' (universals before things) and '*universalia in rebus*' (universals in things) (Price 1953).

2.2.1 Must universals be transcendent? Part I: Russell's arguments

The view that universals are transcendent can be traced to Plato, and is also advanced by Russell in his early work (1912a, Chapters 9–10) who suggests that universals could not be any other way. Each and every blue particular is a spatio-temporally located entity and the universal blue is common to all of them, the essence or common nature of the instances themselves. There is no room in space-time for such multiply-instantiated entities. Put in this way, it is not yet obvious why this should be the case, unless multiple-instantiation itself prevents universals from having spatio-temporal features, but Russell offers two very brief arguments that universals do not exist in space-time. First, he states what one could call the 'argument from immutability': universals do not change, they are not subject to, nor affected by, causality. The particulars which instantiate a specific universal all instantiate a common nature or essence, which 'is not fleeting or changeable like the things of sense: it is eternally itself, immutable and indestructible.' Universals do not change, and so they cannot be spatio-temporally related to anything and thus do not exist in space-time. Now, one might question why being immutable entails not existing in space-time: Why shouldn't space-time contain some entities which do not change?[6]

Although I risk running across a metaphysical minefield, I will sketch two lines of reasoning which might lead one to conclude that space-time could not contain entities which do not change and with which Russell himself might have concurred. The first stems from a specific constitutive, or definitional, view, about the nature of time and space which maintains that space-time is entirely dependent upon changes or causation between entities (and spatial relations between them) and is reducible to them. Furthermore, on this view, entities cannot be spatio-temporally related if they are not involved in causal relations[7] and, if they are not spatio-temporally related, they are not in space-time. Thus, immutable, or causally inert entities are not in space-time. The opposing conception is that of *absolute* space-time, space-time which exists independently of the existence of matter or change and would have existed even if nothing else had. There is no problem for the location of causally inert entities in such space-time since the nature of space-time does not require that the entities located within it can change. In the current context, there is an intermediate position between these two conceptions: space-time which depends for its existence upon the existence of matter and change, but in which immutable entities such as universals can be located too. For instance, immanent universals could have spatio-temporal properties indirectly, perhaps in virtue of the particulars which instantiate them. Neither this latter conception, nor the absolute conception of space-time, preclude the existence of spatio-temporally located universals.

It would require a long metaphysical diversion to thoroughly evaluate the arguments for the conception of space-time required by Russell's view,[8] so I shall be brief. First, it should be noted that were Russell's immutability argument against spatio-temporally located universals to work, it also would rule out other ontological accounts of properties, such as trope theory in which individual instances of properties are fundamental. Tropes do not themselves change but are spatio-temporally located, and so the existence of such entities would be ruled out on the conception of space-time under consideration.

So, why should we accept that the nature of space-time requires entities within it to be causal, or else that all entities within the natural spatio-temporal world just happen to be causally efficacious? Russell might answer that we never experience an entity's having spatio-temporal location except in virtue of its changing; we have no experience of time except through experience of change, and no experience of space except in virtue of the relative positions of particular things. Being multiply exemplifiable and immutable rules out universals on both counts, so they could never be experienced as existing in space-time. But does that absence from our experience entail that such entities are not there?

Russell's second suggested argument, adds that universals cannot be spatio-temporal because all entities in the spatio-temporal world belong to the world of sense we are able to experience their existence and that only particulars can be the objects of sense experience. Both the premises of this argument are contentious: one could deny that all experience is of particulars and say we experience universals after all, or one could challenge the strongly empiricist assumption that the entire contents of space-time are accessible to our senses. First, we have already seen in 2.1.1 how the former claim about the particularity of experience results in difficulties for the empiricist about the meaning of general terms – whenever I experience a tree, I experience a *particular* tree, so how I can understand 'tree' from experience alone is not obvious – a problem which formed the basis of one of the arguments for the existence of universals. Perhaps one could just deny the premise that the objects of individual experiences are particular things. It is not too implausible to say that the colours, and shapes of each of the leaves of the tree, the texture of its bark, its smell, the widest circumference of its trunk, the chemical constituents of the sap and so on are universals, each of which could belong to another tree (or another particular entirely) at another point in space and time. This would not weaken the argument in 2.1.1, since the semantics of general terms would still be explained by universals. It would also make universals more empirically or naturalistically acceptable;

that is, it would allow them to fit in better with the worldview which treats the methods of science, and especially a posteriori observation, as the primary way in which we learn about the world. Of course, one could probably never know that one was experiencing the universal *green* rather than individual greens (say). Furthermore, such a view would need refinement: the senses are usually taken to operate causally – as is the acquisition of knowledge more generally – and so one would need to find a place for universals in the causal order to be consistent with these views. Perhaps one could maintain that universals are not causally inert, or else they could gain a place in the causal order indirectly as *constituents* of causes and effects, such as in Armstrong's view that a cause or an effect is a *state-of-affairs*, a particular instantiating a universal (1997: 14.3).

However, the universals theorist does not require the claim that we experience universals directly, since he could also deny Russell's assertion that all entities in space-time belong to the world of sense. The claim that they do is a significant, and strong, empiricist assumption about the relationship between what exists and our knowledge of it – nothing exists of which we cannot, in principle, have sensory awareness – and it is open to many criticisms. In particular, one might question whether the assumption is coherent and whether it compels us to deny the existence of entities which empirical science seems happy to countenance, such as quarks and black holes. This latter point is the most significant in the present context: our scientific theories appear to tell us that there are some entities which we cannot in principle observe but, while the strong empiricist claims to be the champion of scientific enquiry, according to his theory of what there is such entities do not exist. If we take the side of science rather than that of strong empiricism to find a way out of the conflict, then the existence of entities in space-time of which we cannot have sensory experience is not so peculiar after all.[9] Thus, empiricist-friendly contemporary naturalist philosophers might weaken the empiricist constraint and simply require that our metaphysical hypotheses about what exists in space-time be consistent with those we can empirically test. On the other hand, some realist metaphysicians may treat ontological theorizing as prior to science and place no constraints upon which categories of entities exist on the basis of our epistemic access to them.

2.2.2 Two kinds of strangeness

Russell's arguments that universals do not belong in space-time are not particularly compelling, so the way remains open for the supporter of immanent universals to present his view. Nevertheless, nothing which has been said against Russell's arguments entails that transcendent universals are inconsistent or incoherent, simply that one does not have to accept that universals are transcendent, so it will be instructive to investigate the plausibility of this view as well.

Transcendent universals do not exist in space-time like the ordinary particulars with which we are familiar, like tables and chair and trees and dogs, nor are they entities in space-time which we do not experience directly like electrons, or quarks, or black holes. So where do they exist? And how? Of course, the first question is misguided: it is not simply that transcendent universals are not spatio-temporally located, but that they are not the kind of entities which even could be located; they are *aspatial* and *atemporal,* and so even saying that they are 'nowhere' does not really make sense. However, there is a difference of metaphysical opinion about what we should say about transcendent universals at this point. Russell maintained that such universals '*subsist*', whereas particulars exist. But, as Quine (1948) pointed out, it is controversial to draw a distinction between different modes of being in this way: What does it mean to deny that something exists and yet admit that it subsists? In Quine's view, rather than trying to characterize what subsistence is in opposition to existence, the advocate of transcendent universals should say that transcendent universals exist in the same way as spatio-temporal particulars do, but that universals do not exist in space-time: they are *abstract* (i.e. non-spatio-temporal) entities, as some people also consider numbers to be.[10]

There is a lot of intuitive resistance to transcendent universals: first, their lack of spatio-temporal location makes transcendent universals seem odd (and if they share these properties with other abstract entities then such entities are odd too). It is difficult to conceive of something which has no place, and our intuitive need to place them somewhere makes us imagine them in something like a 'Platonic heaven' where all the abstract objects are. Moreover, one might be mystified about how the relation of instantiation is supposed to work between particulars and universals on the transcendent view: a particular instantiates a universal but whereas the former is a spatio-temporal entity, the latter is not. It is difficult to conceive of entities in space-time as having relations with entities which are not in space-time: witness, for example, the problems which beset substance dualism about how best to characterize interactions between matter and the immaterial mind, because the latter lacks spatial (and perhaps also temporal) location.

However, the supporter of transcendent universals will just accuse his opponents of prejudice: it is just a brute fact that instantiation connects entities in one realm to those of another; that is just what it means for a particular to instantiate a universal and unlike the case of causality, this is not a relation where the relata – the entities related by the relation – seem to require spatio-temporal location. Since we do not yet have a clear idea about the nature of instantiation (which will be discussed in 2.4), this claim seems reasonable, and the intuitive judgements above are not logical arguments against transcendent universals (nor against abstract objects more generally); there are many philosophers who do not share the aversion to entities which do not exist in space-time. Nevertheless, the supporter of immanent universals argues, we would be better placed to understand universals if they were located in the spatio-temporal world as immanent universals are and the simpler theory is usually to be preferred over the more convoluted one. So, is the account of immanent universals simpler and easier to understand?

The conception of universals as immanent restricts the qualities we experience around us firmly to the spatio-temporal world. There is nothing 'beyond' space-time which is more fundamental than the instances of *blue*, or of *heat*, or of *chocolate* which we encounter in our ordinary experience. Furthermore, such universals are not simply multiply-exemplified – that is, repeatable in different parts of time and space – they are also *wholly present* at each place they are instantiated (Armstrong 1989a: 98–9). While a particular cake (say) can only exist in one region of space-time (and completely fills it, to the exclusion of other particulars), the universal *cake* is wholly present in each spatio-temporal region taken up by every instance of a cake that there is, has been, or ever will be. (Of course, many other universals may also be instantiated (and wholly present) in each of the same regions of space-time as those occupied by the universal *cake*.) Moreover, the destruction of one or more instances of a universal – the eating of a cake or two, in this instance, or the destruction of every patisserie in Paris – does nothing to diminish the universal *cake* itself, no part of the universal is destroyed.

Some philosophers have complained that it is not clear how we are to understand this idea of an entity's being wholly present at distinct points in space and time: the whole of something is over here and yet the whole of it is also over there and in many other places simultaneously (Sider (1995: 36-78); Lowe (2006: 24); Heil (2012: 110); Giberman (MS)). Lowe argues that even having accepted that we should not try to understand the spatio-temporal properties of universals as we do those of particulars entities, it is not clear whether we know what 'being wholly present at multiple locations simultaneously' amounts to. In fact he goes further, arguing that the relation of *being wholly present at* leads to absurd consequences, or else we really do not understand it. To do so, he uses the following example: take two tomatoes, b and c, both of which instantiate the same shade of redness universal R. It is plausible that the relation *being wholly in the same place as* is an equivalence relation and therefore that it is both symmetric and transitive. Now, b instantiates R, so R *is wholly in the same place as* b, and c instantiates R so R *is wholly in the same place as* c; by symmetry b *is wholly in the same place as* R, so by transitivity b *is wholly in the same place as* c which is absurd.

The distinct particular tomatoes b and c should not turn out to be wholly in the same place as each other. The obvious response is to deny the symmetry of the relation: R *is wholly in the same place as* b, but b *is not wholly in the same place as* R. This would block the unwanted conclusion, but it makes the notion of universals being wholly present in the same spatio-temporal region as the particulars which instantiate them even more mysterious than before. (Lowe calls this response 'unprincipled and question-begging' and diagnoses the problem as being the result of conflating universals with their instances, which he takes to be individuals – property-instances or tropes – which will be discussed in Chapter 3.) Armstrong describes particulars as a qualitative 'layer-cake', with instances of different immanent universals (that is, the whole universals themselves) being wholly located in their respective layers (1980: 108). But although this metaphor permits us to deny that universals are co-located with the particulars which instantiate them (each universal is just wholly located within a layer of that particular) and this avoids Lowe's problem, it requires the 'extension' of particulars along a *qualitative* dimension which ordinary space-time certainly does not appear to provide. We have swapped the strange non-location of transcendent universals for the strange multiple locations of immanent universals and it is not obvious that the latter is any more plausible than the former.

2.3 Must universals be transcendent? Part II: Ideals, aliens and problems with resemblance

Above, we considered arguments for transcendent universals on the basis of their immutability or their supposedly being inaccessible to sensory experience and found these both to be inconclusive reasons for thinking that universals cannot exist in space-time. Now we will examine three more respects in which the doctrine of transcendent universals claims to have an advantage.

2.3.1 Ideal standards

The first group of examples concerns the existence of universals which are never exactly, or completely instantiated: *perfect beauty*, or *perfect courage*, or the *perfect circle*, are never instantiated by spatio-temporal particulars but exist as abstract entities, as the common natures or limits to which imperfect, worldly instances approximate. (Even if one has qualms about the objective existence of universals for ethical, political and aesthetic qualities and has already rejected these for independent reasons, the problematic examples of perfect mathematical universals would still remain.) There are many instances of circles, but each one is flawed and inaccurate in some way or other; the entity that conforms exactly to the mathematical properties of a circle does not exist in space-time. If one takes the immanent view of universals then the nature of a universal is exhausted by its instances and there are no extra ontological resources with which to explain these kinds of imperfect match to an ideal standard; that is, why a circle is somehow more than the instances which approximate it. Moreover, the doctrine of immanent universals is bound by the *Principle of Instantiation*: universals with no instances do not exist, and so there can be no ideal standards such as the perfect circle, or the perfectly straight line, in virtue of which the imperfect instances might resemble each other and to which each instance approximates to a greater or lesser degree.

In response to this objection, the supporter of immanent universals can simply accept this conclusion and deny the existence of a range of universals which are never instantiated, such as *perfect courage*, *perfect beauty* and mathematical entities such as the *perfect circle* or the *perfectly straight line*. According to this theory, these entities are ideas or fictions, not real universals which exist independently of human thought. But is it plausible to characterize a range of mathematical entities in this way? David Armstrong, one of the most vociferous advocates of immanent universals, maintains that it is: we can extrapolate from our experiences of particular circles which approximate more or less to an ideal standard, and form an idea of something which does meet the standard, the perfect circle. 'Why attribute metaphysical reality to such standards?' he asks (1989: 68). The answer from his opponents would be that they do not think that an idea, or a fiction, can do the work we expect of a mathematical entity. Geometry for Armstrong is either about the properties and relations of imperfect entities in the spatio-temporal world, or it is about the properties of mind-dependent entities. But for many philosophers of mathematics, the subject matter of geometry and mathematics more generally is independent of both the human mind and the physical world and if that is so, universals must be transcendent in their account.

2.3.2 *Uninstantiated universals*

A second reason for thinking that transcendent universals are required in metaphysics is based upon the alleged existence of uninstantiated, or *alien* universals. While transcendent universals can exist which have no instances, the range of immanent universals is restricted to those which have at least one instance. This restriction is known as the *Principle of Instantiation* which, to make it more plausible, is treated as ranging over the whole of time; it is enough for the existence of a specific immanent universal for it to have been instantiated at least once, or that it will be at some time in the future. It is worth noting that the entailment does not hold in the other direction – the Principle of Instantiation does not entail that universals are immanent – and so there is nothing contradictory about characterizing universals as transcendent, abstract entities and yet requiring that each has at least one instance (Lowe 2006: 25). Thus, the Principle of Instantiation does not mark out the same divide as that between transcendent and immanent universals.

The current difficulty is that the proponent of immanent universals is committed to the claim that there are no universals which could have had instances but do not actually have them, and yet there are several counterexamples to this view. If such counterexamples cannot be explained away, the Principle of Instantiation is too restrictive and there are not enough universals on the immanent view. The Principle of Instantiation has already created similar problems above concerning the existence of ideal standards, universals which have no instances that match them exactly, such as the perfect circle or the perfectly straight line, examples which the supporter of immanent universals has already attempted to explain away. But one might think that there are also universals which are not even approximated by any particular, and yet it seems accidental or contingent that this is so: a shape with a google+23 sides (that's $10^{100}+23$), or an 'alien' physical property which makes gravitational forces weaken less rapidly as the distance between masses becomes greater than they actually do. Similarly, it seems to be contingent that there are no creatures such as dragons (even the biological mechanism for fire-breathing does not seem so unlikely as to be physically impossible). Even more serious omissions suggest themselves, such as the universals associated with the higher elements of the periodic table, elements which do not occur naturally and in some cases have been created in laboratory conditions. The existence of universals for elements with atomic number 119 and above is entirely dependent upon our ability to create them, which we have not yet done. If we manage to do so, such universals exist now; if we give up chemistry and physics tomorrow and no other creature in the universe steps in to do the work, they do not exist. Moreover, until recently there had been no accredited success in creating element 113 (ununtrium) (at the time of writing, its 'discovery' still awaits confirmation). If confirmation is refused and we never do succeed in providing an instance of it, there will be universals for each element around element 113 and no element 113, even though we have a good idea of the chemical and physical behaviour that element 113 would exhibit, and its lack of instances seems to be down to human abilities. The Principle of Instantiation seems too restrictive to be plausible.

What is the proponent of immanent universals going to say in such cases? Should he abandon the Principle of Instantiation and accept that some trans-cendent universals exist? If he does so, he might as well accept that *all* universals are transcendent: it seems ad hoc to say that those which actually have instances are immanent, and those which could have had instances are transcendent.

David Armstrong takes an uncompromising line in the face of such examples:

> For myself, I do not see the force of the argument [for the existence of uninstantiated universals]. Philosophers do not reason that way about particulars. They do not argue that it is empirically possible that present-day France should be a monarchy and therefore that the present king of France exists, although, unfortunately for French royalists, he is not instantiated. Why argue in the same way about universals? Is it that philosophers think that the universals are so special that they can exist whether or not particular things, which are contingent only, exist? If so, I think that this is no better than a prejudice, perhaps inherited from Plato. (1989a: 69)

Is this attitude justified? The implausibility of some of the contingent scientific 'gaps' in the range of universals discussed above suggests that Armstrong's answer is too glib, but his hard line, parsimonious stance is consistent: if instantiation is required for existence then, despite appear-ances and intuitions to the contrary, there is no element 113 if nobody manages to create it.

One solution which the immanent universals theorist could adopt would be to reject the actualism which has been implicit in the Principle of Instantiation and to accept modal realism instead. This would involve rejecting the view that what is possible is determined by entities existing in the actual world and accepting that possible entities exist in the same sense as actual ones (Lewis 1986). If we reify possible worlds and their contents in this way, there is nothing to stop the alien universals which are contingently 'missing' from the actual world being instantiated by possible particulars. One can provide sufficient immanent universals if their instantiation is not restricted to the space-time we actually find ourselves in. However, there are three points to note about this potential solution: first, it would be at odds with Armstrong himself who is committed to actualism and for whom the Principle of Instantiation requires that universals be instantiated in the space-time we are in; second, it will not solve the other difficulties with immanent universals concerning ideal standards[11] and degrees of resemblance which are discussed in 2.3.3; and third, it involves a much higher degree of ontological commitment than the actualist account of immanent universals, so that one of the major advantages of the immanent conception over the transcendent one would be lost.

Problematic scientific examples aside, the question of which side one should take in this debate is partially governed by what universals are thought to do, the wider philosophical role which one takes them to play. Hints of this disagreement are contained within the passage quoted from Armstrong above, although it's not clear whether he realizes it. For those who think universals are the referents of predicates, the existence of a universal which is the referent for the predicate 'is a present King of France' is an attractive option. This might initially sound strange, but such a move permits one to give an account of the meaning of statements about non-existent 'things' (the meaning of 'empty names' as they are often called, or singular terms which lack referents) without having to postulate a realm of *particulars* which have the property of non-existence[12] (Russell 1905).

In some respects, the supporters of transcendent and immanent universals are arguing at cross-purposes: the supporter of transcendent universals thinks that his opponent has too few universals, and indeed he does for what the transcendent universals theorist wants universals for; while the proponent of immanent universals thinks that his opponent has too many universals, which is (probably) true if one is predominantly concerned with the behaviour and categories of the actual natural world. These competing aims also create areas of apparent conflict for other property theories too and eventually result in different conceptions of properties which are not thought to compete (and, which on some accounts, are related to each other). Since this problem is not specific to the theory of universals, I will delay discussion of it until Chapter 5.

2.3.3 *Problems with resemblance*

So far, the immanent universals theorist has stood his ground in the face of counterexamples to his theory, but there are two interrelated difficulties which are harder to block in this way: the first concerns cases of *inexact* resemblance between particulars; and the second how he should deal with similarities between universals themselves.

The first problem arises because some universals have seemingly disparate collections of instances. For example, a patch of International Klein Blue paint, Elvis Presley's suede shoes, a particular mountain bluebird, and my broken umbrella all instantiate the universal *blue*, but they do not instantiate the same shade of blue in each case. Likewise, every instance of courage differs from the others in the actions performed, or the people involved, or the circumstances of their actions, but all instantiate the universal *courage*. The particulars in these examples resemble each other, but they do not resemble *exactly* and it is not clear how we can explain this inexact resemblance in terms of immanent universals.

Such variation between instances can be explained on the transcendent view because universals exist independently of the particulars which instantiate them and there need not be a discernible common nature to the instances in virtue of which the universal is instantiated. Moreover, different particulars can instantiate a transcendent universal to a greater or lesser degree: it is a little bit courageous for a timid student to stand up and demonstrate to a crowded lecture theatre why my argument is flawed; whereas her facing certain death to save everyone in the room instantiates a far greater degree of courage.

Exact qualitative similarity is accounted for by identity (on both accounts of universals) but it is not yet clear how the immanent universals theory is to account for inexact similarity; how can the same universal be wholly present in each of the inexactly resembling instances? A natural answer suggests itself from the one-over-many argument: many different instances of blue are actually instances of different universal shades of blue, which are themselves similar in virtue of their instantiating the universal *blue* as well, perhaps in virtue of their sharing some other properties. Furthermore, blue things generally inexactly resemble green things, and those of any other colour, because they instantiate the universal *colour*. But if the immanent universals theorist offers this as a solution, he has merely deferred his difficulties, since now we might reasonably ask *why* green, blue and orange resemble each other, and also why blue resembles green more than it does orange. We now have a problem about how we can explain the inexact resemblances between *universals* themselves. This additional problem arises more rapidly in cases where there is no common set of universals which each particular instance instantiates and yet a family resemblance obtains between the particulars, such as in Wittgenstein's example of the entities to which the word 'game' applies. To illustrate the increasing difficulty of the examples with which the immanent universals theorist is faced, consider the schema on Table 2.1 (borrowed from Armstrong 1989a: 85):

Table 2.1

Particulars	a	b	c	d	e
Their properties:	F G H J	G H J K	H J K L	J K L M	K L M N

Table 2.1 shows particulars with overlapping groups of properties which do not instantiate universals in common, such as might be found in a collection of particulars to which the word 'game' applies. This is intended to explain the family resemblance or similarity between the cases a – e, and thus why the term 'game' applies without requiring a (transcendent) universal *game* which a – e all instantiate despite their lacking a common nature. But, as Armstrong later admits, the situation is not as simple as this for the immanent universals theorist. Let us say that H is mass and J is colour (and the subscripts indicate different masses and colours), then the more accurate picture is something like that in Table 2.2:

Table 2.2

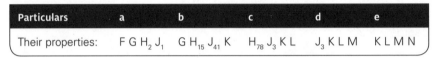

Particulars	a	b	c	d	e
Their properties:	$F\,G\,H_2\,J_1$	$G\,H_{15}\,J_{41}\,K$	$H_{78}\,J_3\,K\,L$	$J_3\,K\,L\,M$	$K\,L\,M\,N$

Here, there is no overlap between particulars a and c (say, for instance, that a is blue with a mass of 1.4 kg, while c is black-and-white with a mass of 0.4 kg) unless we have an account of why the respective instances of mass and colour resemble each other. Armstrong's account of 'likeness without identity' (from the first diagram) by which he hopes to explain problem cases of inexact resemblance is too simple, it is hampered by the sheer prevalence of likeness without identity.

It seems that the immanent universals theorist will have to resort to postulating higher order universals again: *blue*, which is common to shades of blue; *colour*, which is common to blue, green and red; *mass*, which is common to 1 kg, 0.23 g, 14 ounces and 117 tonnes. One could, at this point, suggest that the immanent universals theorist draws upon the distinction between *determinate* and *determinable* universals: the universal *colour* is a determinable, while *blue* and *green* are determinates of it; or 1 kg and 14 ounces, are determinates of the determinable *mass* (Price 1953: 36–8). However, it is not obvious how this resolves the problem because in all but the simplest cases, what counts as a determinable depends upon what perspective we take on a more extensive ordering: from one perspective, *blue* and *green* are determinates of the determinable *colour*, from another, the particular shades of blue are determinate cases of determinable *blue*. The determinate/determinable distinction describes a relationship between universals at each stage in the order, but no more. Furthermore, the distinction does not really seem to *explain* anything about the resemblances between the universals of increasingly higher orders which we have postulated: Why do universals fall into such orders at all?

In addition, postulating higher orders of immanent universals threatens the parsimony of the theory: higher-order immanent universals are all instantiated by a particular whenever those of lower orders are. I thought my broken umbrella was just blue but it turns out to be *cobalt blue, blue, blue-green, coloured* and perhaps many other colour-classifying universals besides, the existence of which I only know about because I see the umbrella as blue. Unlimited higher order universals are epistemically unprincipled and uneconomical from the point of view of a theory of immanent universals (although this abundance would not be problematic for an account of transcendent universals, which does not need them anyway to explain inexact resemblance), and Armstrong is rightly wary of introducing them into his theory. But what are the alternatives?

Armstrong suggests that while resemblances between simple universals involve a higher order universal (the determinate/determinable cases), most resemblance between universals is between complex universals which can be treated as 'overlapping' each other. For example, a particular with a mass of 3 kg can be distinguished into parts with mass of 2 kg and 1 kg (say); thus there is a clear sense in which we can say that 2 kg resembles 3 kg, since any particular instantiating the latter has a 2 kg part. I will not examine Armstrong's claim in detail here. In some cases, he is forced to rely upon the notion of some universals, such as colour, as having a concealed complexity which we are nevertheless aware of through its operations on us (because we see the resemblance). This complexity then allows us to tell a similar story about colour resemblance as we did about mass, but the fact that it can be both concealed and we can be aware of it is obscure and its postulation seems ad hoc. (See Armstrong 1988; and Pautz 1997 for a short but technical criticism of his position.) Armstrong's account is too vague and it contains some substantial assumptions about the natural world, so the prospects of his strategy succeeding are unclear. If it fails, the universals theorist is left with higher orders of universals but no explanation about why universals fall into those orders. The remaining option is to postulate *primitive* resemblance relationships between universals; that is, to claim that some universals resemble each other and fall into orders but that resemblance *cannot* be more fully explained. If the immanent universals theorist takes this route, he has conceded that his universals theory needs primitive, unanalysable resemblance just as much as some rival ontological accounts of properties such as trope theory do (3.2). Why not, the trope theorist might ask, just cut out the uneconomical and spatially counterintuitive immanent universals and give an ontological account of properties in terms of their individual instances instead?

2.4 Problems with instantiation: The instantiation 'regress'

A particular object is a dog in virtue of instantiating the universal *dog*, a picture is blue in virtue of instantiating the universal *blue*; but, what is *instantiation*? So far, although I have talked about particulars instantiating universals, I have had very little to say about what links particulars to universals, what the relationship between a universal and the particulars which instantiate it amounts to. This might turn out to be an important omission, since some philosophers have suggested that *instantiation* leads to an infinite regress and thereby makes the whole account of universals untenable. This problem is known as Bradley's Regress (although it is not clear how much Bradley (1893) had this problem in mind), the nexus regress, or the relation regress; and we have already encountered another variant of it in the guise of the Resemblance Regress (2.1.2), in which it was used to argue *for* universals against other ontological views, in that case against the account of properties in terms of resemblances between particulars. Obviously, if universals theory is susceptible to an analogous regress, the plausibility of using the Resemblance Regress as an argument in favour of universals will be seriously damaged.

The regress argument applies to accounts of universals in the following way. Let particular *a* instantiate universal F: *a* instantiates F. *a* can instantiate other universals (G, H and so on), while other particulars (b, c, d, etc.) can instantiate F (and other universals too), so the instantiation which binds *a* to F is a multiply-exemplified relation; *instantiation* is itself a universal. The instantiation relation which links *a* and F is an instance of a universal relation I_1 (say) holding between *a* and F. But now, F *and a* instantiate the instantiation relation I_1 (F and *a* *instantiate* I_1); and so there must be a three-place relation of instantiation I_2 which links F, *a* and I_1. But then, F, *a* and I_1 instantiate I_2 and we require a four-place relation I_3 to connect F, *a*, I_1 and I_2. Clearly, this is not going to stop. Each case of instantiation presupposes the existence of another instantiation relation with a higher number of argument places I_1, I_2, I_3 and so on which are not prima facie the same relation as any of the previous ones, since each relates a different number of entities. If the regress is vicious, then nothing can instantiate anything else and both the accounts of universals which we have been discussing are in danger.

The Instantiation Regress

a is F
What connects F to a?   universal 2-place relation I_1:   x instantiates$_1$ y
What connects F, a, and I_1?   universal 3-place relation I_2:
                                x and y instantiate$_2$ z
What connects F, a, I_1 and I_2? universal 4-place relation I_3:
                                x and y and z instantiate$_3$ x$_1$
And so on, ad infinitum.

Each stage requires the existence of another universal instantiation relation to link the existing components and so at no point are the universals and the particular bound to each other.

Two principal strategies are employed against this objection: first, denying that the regress is vicious and accommodating the infinitude of instantiation relations; or, second, accepting that an instantiation relation would lead to regress while denying that instantiation is a relation, thereby dodging the damaging conclusion. I will discuss these responses in turn.

2.4.1 The 'regress is not vicious' response

The first strategy relies upon making a crucial distinction between different ways of viewing the original problem: an *internalist* reading and an *externalist* one (Orilia 2006). The internalist reading goes as follows: we started out with a single state of affairs, S, with constituents F and a, a particular a instanti-ating a universal F. S turned out to require relation I_1 as an extra constituent in order to link F and a; but, given the addition of I_1 to S (so S now includes a, F and I_1), the characterization of S turns out to require I_2 as an additional constituent too; but now S is constituted by a, F, I_1 and I_2 and so requires I_3 to be included; and so on forever. S *cannot* exist, Orilia argues, because it requires an infinitude of constituents to exist in order to bind it into a unified state of affairs, so the regress is vicious.

The externalist reading of the regress is different: we start out in the same way with a state of affairs Fa, particular a instantiating universal F, let us call this S. Now, the state of affairs S of a's instantiating F instantiates the universal I_1: we require a *new* state of affairs a's instantiating F instantiates I_1, let us call it S_1. S_1 is external to S, and so although S_1 is required for S's existence, it does not make the existence of S impossible. The process described in the externalist reading will also continue ad infinitum: there will be S_2 formed by the three-place relation I_2 which connects F, a, and I_1; and S_3 to characterize the instantiation of relation I_3 between F, a, I_1 and I_2. We have infinitely many instantiation relations producing infinitely many states of affairs, but the regress is not vicious.[13]

On a transcendent conception of universals, one might take the view that this complicated infinite series of universals can be accommodated and the infinitude of states of affairs simply arises as a result of particulars and universals instantiating these higher-order instantiation universals; without commitment to the Principle of Instantiation, there is no prima facie reason why we shouldn't allow that such infinite hierarchies of universals exist. Given infinitude in the natural, spatio-temporal world, the supporter of immanent universals might be able to accept this externalist reading of the regress too, although on pain of requiring a massive over-abundance of entities in the natural world. Where one thought that there was simply a particular a instantiating a universal F, there is an infinite hierarchy of states-of-affairs and that does not seem a very compelling account of instantiation; as in the case of the higher-order immanent universals in the previous section, it is ontologically uneconomical and epistemologically rather suspect, since the only reason we have to postulate these extra states-of-affairs is to avoid the threat of regress.

Furthermore, one might contest the implications which Orilia draws from the distinction between the internalist and externalist readings, in particular his claim that the internalist reading leads to a vicious regress while the externalist one does not. According to Orilia, the externalist reading is unproblematic if one is prepared to countenance infinite hierarchies of state-of-affairs, but the single state-of-affairs proposed by the internalist reading to characterize instantiation cannot exist because it requires infinitely many constituents in order to do so. However, such an infinitely complex state-of-affairs could exist if we were prepared to accept that the world is infinitely complex, such that at least some entities are constituted by infinitely many entities, or matter is infinitely divisible for example. This requirement for infinite complexity differs from the infinitude required on the externalist reading, which 'simply' requires that the world contains infinitely many state-of-affairs (with finite constituents). The latter is much less controversial than the former and so, although one can argue that the internalist reading is coherent – and the states-of-affairs it postulates possible, contrary to Orilia's claim – the fact that it requires the world to be infinitely complex may be considered a drawback, since this places an arbitrary restriction on the ways the world could be.

Perhaps we could attempt to simplify the infinite hierarchy of instantiation relations required by the regress. We have assumed that I_1, I_2 and I_3 are distinct relations; but is it plausible to treat them as being identical? Those familiar with elementary logic might quickly conclude that the answer is 'no' because I_1, I_2 and I_3 are each relations with a different number of argument places, they each relate a different number of entities (2, 3 and 4 in this case), but we can allow multigrade, or variably polyadic, relations which can vary in the number of entities they relate. Thus, there are two options here: either we could postulate a single multigrade relation I of which I_1, I_2 and I_3 are instances; or, we could treat I_1, I_2, I_3 and so on as being inexactly resembling instances of a single instantiation universal I^* (say). The latter option requires a successful account of inexact resemblance between universals (of the type discussed at the end of 2.3.3) but with this in place, we could give an account of I_1, I_2 and I_3 and so on all being instances of I^* in a similar way to the universals *red*, *orange* and *green* all instantiating the universal *colour*. However, whichever of the two overarching instantiation relations is deemed preferable, both I and I^* need to be able to self-instantiate, in order to avoid problems being generated by cases such as I_1 instantiates I^*, I_2 instantiates I^*, which must also be instances of I^*. There might be good reasons for banning self-instantiation as part of a wider ban on universals, or properties in general, exemplifying themselves, but in the absence of an outright ban, these simplifications remain tenable.

2.4.2 *The* 'instantiation is metaphysical glue' *response*

The alternative to attempting the rather complex manoeuvres above is to accept the viciousness of the regress as described but to deny that instantiation is a relation at all. According to this response, instantiation is a fundamental, non-relational tie or nexus which binds or glues a particular to a universal it instantiates (perhaps thereby binding them into a state-of-affairs), it is not a genuine relation at all because its nature is exhausted by its relata, that is, by the particular and the universal themselves. In a similar approach, Broad (1933: 85) likens instantiation between particular and universal to the glue between two sheets of paper: it just sticks them together and is itself self-adhesive, it does not require any additional sticky substance between paper and glue.

The postulation of an internal relation works as a solution to the Resemblance Regress, discussed in 2.1.2: we can treat resemblance as being an internal relation which obtains in virtue of the internal natures of the particulars which it relates, thereby diluting the force of that regress to a trivial logical trick; all that is needed for particulars *b* and *c* to resemble each other is for *b* and *c* to exist. However, the situation is not so simple in the case of instantiation: particular *b* and universal F can both exist and yet *b* not instantiate F; the instantiation relation between particular and universal is contingent and does not obtain in virtue of the internal natures of its relata alone. For instance, we have no account of why particular b is non-relationally tied to universal F, while c is not. The fact that b and F are in the same spatio-temporal region does not help to make the non-relational tie more explicable: presumably F is wholly present where b is *because* b instantiates F and so the shared location of b and F cannot be used to explain why b instantiates F.

Given this disanalogy between instantiation and other versions of the regress, Armstrong argues that we must also accept the existence of a funda-mental tie between particular and universal as primitive: that is, a 'universal-like entity' *instantiation* is required between a and F, but then the combination of a, F and instantiation is sufficient to ground an internal relation which binds the three. Thus, he claims, the regress we can generate, such that 'a, F and *instan-tiation* instantiate instantiation' and so on, is logical not ontological; although we are 'logically forced' to proceed to the next step, the progression does not require the existence of any additional truth-maker nor ontological ground at each stage. The regress has been halted after the first step when *instantiation* was introduced because *instantiation* is an internal relation which requires nothing more than the existence of its relata – in this case, the particular a, F and *instantiation* itself – to hold[14] (1989a: 53-7; 108-10). Internal relations, Armstrong claims, are not ontologically robust entities: they add nothing extra to the furniture of the world, they are an 'ontological free lunch' (1989a: 56) and not the kinds of entities we need focus on in ontology (1997: 92). This would be a neat response, if it is plausible, halting the regress before it really gets started, although it seems more plausible as an explanation of the instan-tiation of immanent universals than for that of their transcendent cousins. In the case of immanent universals, the universal is wholly present in the same spatio-temporal region as the particular, so there is no peculiar non-relational 'action at a distance' involved in the primitive notion of instantiation required. But when a particular instantiates a transcendent universal, the particular is in space-time while the universal is not, with instantiation somehow connecting the two: How could there be a *non*-relational connection between entities of the abstract and concrete realms? Or, even an internal relation between them? (As noted above, it is hard enough to make sense of a relational connection between abstract entities and spatio-temporally located ones.)

One broadly Fregean suggestion is that transcendent universals are *unsaturated*; that is, roughly speaking, universals essentially have gaps which can be filled by particulars and the filling of those gaps does not require any additional ontological mechanism such as instantiation. (We can draw a direct analogy with predicates and the objects needed to satisfy them here: the predicate '... is wise' is like a sentence with a gap in it which can be filled with a subject such as 'Socrates'. We do not have to explain why 'Socrates' fits in the gap to give the sentence 'Socrates is wise' and we may well be forced into a regress if we try.) The difficulty with this solution, aside from the fact it is rather vague and metaphorical, is that it does not explain *why* the gap in a universal F is filled by a specific particular *b* and not by a different particular *c*; this account of the relationship between universals and particulars is too general to explain specific cases.

Though these are popular solutions to the regress, one might be concerned that we do not really understand the metaphors involved: What is this primitive, in principle inexplicable, non-relational tie of instantiation? How does fundamental *instantiation* for Armstrong manage to be 'universal-like' while not a universal? What, on the Fregean account, are the 'gaps' in universals? While any theory has to take some terms as primitives and to accept some assumptions without argument, one is left with the nagging feeling that this family of responses to the instantiation regress simply legislate their way out of the problem. However, perhaps this is not as unprincipled as it sounds: if competing accounts of the ontology of properties have to adopt analogous measures to avoid a vicious regress, perhaps by adding primitive features to their ontology, then the theories of universals is no worse off than them.

2.5 A puzzle about self-instantiation

Can universals instantiate themselves? It would seem that the universal *universal* does just that, and one of the responses to the instantiation regress relied on the possibility of self-instantiation to make it work (2.4.1). But some philosophers have argued that there should be a ban on universals self-instan-tiating, or self-exemplifying as it is sometimes called, such that no universals can exist which are instances of themselves. Let us briefly explore why such a restriction might be considered necessary and, since it was explicitly removed in an attempt to solve the instantiation regress above, whether that removal is always illegitimate.

For a difficulty to arise about self-instantiation, two features need to be true of an account of universals (or of properties more generally):

(i)   every universal F has an associated complement universal *not* F;

(ii)   there is a universal relation *self-instantiation*, as there would be if universals can always or sometimes self-instantiate, or the universal relation *instantiation* is simply allowed to be reflexive.

If these two features are true of the theory of universals, as they would be for example if every predicate were associated with a universal, then there is a universal for all the things which are *non-self-instantiating*, everything which does not instantiate itself. This turns out to be paradoxical when we consider what happens to this universal *non-self-instantiating* itself: if it does not instantiate itself, then it turns out that it is non-self-instantiating and so it does instantiate itself; in which case, it does not instantiate itself after all; and so it does self-instantiate. (It is normal to be confused at this point!) Such a universal cannot exist and so the two features above are not consistent.

This paradox – a version of Russell's paradox – led to some philosophers, including Russell himself, introducing a complete ban on self-instantiation. This produces an extra dimension of complexity to the theory by dividing universals, or properties, into a hierarchy of types, in addition to having questionable consequences such as banning the universal *universal* from self-instantiating although this self-instantiation is harmless enough.[15] Perhaps, as an alternative to an outright ban, one could reject claim (i) and deny that there are negative or complement universals *not*-F for each universal F, although this is not an uncontroversial move. If this is not satisfactory, Russell's solution might be in order, but this will rule out some solutions to the instantiation regress in 2.4.1. The universals theorist needs to take care about self-instantiation and its employment in conjunction with other assumptions of his theory, but it does not have to be ruled out completely.[16]

2.6 Conclusions about universals

Let us briefly summarize some key points from this chapter in the hunt for some interim metaphysical conclusions. First, the arguments for the existence of universals are not deductive arguments as such but intuitive nudges in the direction of accepting the existence of universals for semantic or ontological reasons. (The strongest of these, Russell's argument that universals are required by other accounts of properties, fails at the point that universals theories are faced with a regress too.) Treating exact qualitative similarity as identity provides a simple and perspicuous explanation of the ontology of properties, on condition that resolving the questions covered later about location and instantiation does not complicate the whole picture and forfeit this advantage.

The plausibility of transcendent universals is going to depend in part on how plausible one finds abstract entities to be and also upon whether one thinks that there are more universals than the actual world can provide (which in turn may depend upon the ontological role in semantics or modality that one takes universals to play). On the other hand, the question of whether universals can be treated as immanent depends upon the resolution of three main difficulties:

the plausibility of the rather peculiar spatio-temporal feature of a universal's being wholly present wherever each of its instances is located; whether it is coherent to suppose that there are no uninstantiated universals; and whether the account of inexact resemblance between immanent universals can be made to work. Finally, both accounts of universals have problems with instantiation, which their proponents may try to solve by accepting infinite complexity (most plausibly in the case of transcendent universals) or simply by taking instantiation to be non-relational and fundamental.

This is not the time to make a final decision between these ontological theories, since the alternatives to universals have not yet been discussed. Nevertheless, on considering the problems faced by the account of immanent universals concerning location and inexact resemblance, one might begin to think that a more plausible ontological account of properties based on those in the actual world might be developed in terms of the individual property-instances or tropes themselves, without the additional category of immanent universals to complicate matters. It will be to trope theories that I turn in the next chapter.

· FURTHER READING

Russell 1912a; Armstrong 1978a or 1989a; Lowe 2006.
Lewis 1983a; Ramsey 1925.

Suggested Questions

1 Do two triangles have something in common? If so, what is it?
2 Where are universals and why?
3 What is the relationship between a universal and its instances?
4 Compare and contrast the instantiation regress with the resemblance regress. Is either one more harmful than the other?
5 Are there good reasons for thinking that there are any uninstantiated universals? What are the implications of your answer for the theory of universals?
6 What are the advantages and disadvantages of a theory of universals?

Notes

1 Lowe (2006) distinguishes instantiation, which holds between particular substances/things and substantial universals, or property-instances (tropes) and non-substantial universals, while exemplification is an indirect relationship between particular substances and non-substantial universals

(2006: 20–3). Since I will not be considering how or whether to combine universals with tropes, I will not honour Lowe's distinctions here.

2   *Republic*, 596a. Translations of this sentence vary, with some translators preferring the more definite 'You know that we always ...' at the start, but I will not delve into the controversies here.

3   The question of whether we can gain understanding of general terms from our experiences of particular entities was a matter of lively debate between Locke and Berkeley.

4   See especially Armstrong 1978a. Some of his objections will be presented later when the accounts of properties to which they apply are discussed, so I will not repeat them here. (3.2, 9.5)

5   The situation is more complicated still, since the orders are not neatly divided, contrary to what I have presumed: each higher-order ordered pair must also be related to all those in the orders below it as well as those in the same order. For instance, <a, b> resembles <<a,b>, <a, c>>, giving the pairs <<a,b>, <<a,b>, <a, c>>> and <<<a,b>, <a, c>>, <a, b>>, and so on.

6   The converse *is* plausible: that entities which are subject to change, or stand in causal relations to each other must also be spatio-temporally related. We usually conceive of cause and effect such that the former occurs *before* the latter and the cause occurs in a contiguous spatial region to the effect (or at least that an intermediate causal chain exists next to the effect). To deny these would be to countenance 'action at a distance', a phenomenon which seems to make causal interactions mysterious indeed. Thus there is a genuine metaphysical puzzle about how entities which lack spatio-temporal location can be causally efficacious (see, for example, Benacerraf's (1973) problem in the philosophy of mathematics: If numbers are not spatio-temporal and we accept a causal theory of knowledge, then how can we have mathematical knowledge?).

7   One might prefer the term 'causal relations' with scare quotes around it here, as Russell was no fan of causality or causal relations (1912b). Those who wish can just talk about 'changes'.

8   See Shoemaker 1969, Newton-Smith 1980.

9   One can bring empiricism back in line with science and give a different story about the problematic entities, but I will not explore that option here (see, for example, van Fraassen 1980).

10   See, for instance, Frege 1884. Quine himself was no fan of universals, nor properties either and he was a fairly reluctant exponent of abstract objects: his target was simply the coherence of making a distinction between ways in which entities exist. Russell, on the other hand, had his own reasons for distinguishing subsistence from existence and his insistence that transcendent universals should be treated as subsisting. Russell's position can at least partially be traced to his rejecting Meinong's thesis (1904) that there are subsisting *particulars* as well as existing ones. See Russell 1905.

11   One might try to argue that it is possible for the ideal standards to be instantiated by spatio-temporal particulars in another possible world

and thus that they would be accounted for by a modal realist theory of immanent universals. However, this solution rather misses the point of what the ideal standards involve.

12 This does not preclude giving an account of meaning in terms of immanent universals, but the relationship between the universals which there are and predicate meaning will have to be more complicated to deal with all the predicates there might be. See Mellor 1993.

13 See Gaskin (2008, Chapter 6) for a version of this solution who calls it 'an innocent, constituting regress' (2008: 331).

14 This attempt to dilute the regress is not dissimilar to the attempt at the end of the previous section to treat the infinitude of instantiation relations as one multigrade relation **I**, or as being inexactly-resembling instances of one universal I*.

15 A discussion of Russell's theory goes outside the scope of this chapter.

16 Similar notions such as self-predication led Plato into some interesting difficulties with his own theory. See *Parmenides* and also Cohen 1971.

3

Tropes

Different formulations of trope theory are explored in which properties are treated as unrepeatable individual qualities. Three accounts of resemblance between tropes are considered: trope resemblance as a primitive fact; resemblance relations obtaining between tropes; or tropes falling into natural classes. Objections are raised to the latter two: the first leads to regress; and the latter gives a counterintuitive account of the nature of a trope. The primitivist account also encounters a serious problem however, as consistency with truth-maker theory requires tropes to manifest the dual aspects of particularity and qualitative nature which (if they are to be regarded as simple entities) they cannot do. Responses to this objection are found to be less than compelling; but it could be sidestepped by providing an external account of the individuation of tropes. Individuation conditions for tropes are discussed, and then arguments for trope theory surveyed.

In 1957, an exhibition opened at the Galleria Appollinaire in Milan containing eleven paintings by the French artist Yves Klein, each one a monochrome canvas measuring 77.5 x 55.2 cm, each one entirely painted in the blue shade which the artist later patented as 'International Klein Blue' and yet each on sale at a different price on the basis that each painting presented 'a completely different essence and atmosphere' (quoted in Banai 2004: 19). The philosophical point of this example is that despite the qualitative similarity between the colour, shape, size and texture of the paintings, Klein maintained that each one has individual qualities distinct from the others, and this observation about the particularity of property instances takes us in the direction of trope theory.

Let us broaden this observation and suppose that each occurrence of blue is a particular entity: the blue of the clear sky this morning, of each one of Elvis Presley's blue suede shoes, the International Klein Blue paint on each of the eleven monochrome paintings in Yves Klein's 1957 Milan exhibition, the surface of a later painting IKB79 in the Tate collection, and each spot of IKB paint spilt while painting it, are all individual, unrepeatable qualities. Similarly, the qualities of an individual dog – Pharaoh, the Great Dane cross, for example – are all individual entities: his shape at a particular time, his mass, his genetic code, his black colour, his smell are each particular qualities co-located within one dog-shaped spatio-temporal region.[1] From one perspective, the individual qualities which make up this particular dog are no more respectively related to other particular dog-shapes, blacks, masses of 88 kg, genetic codes, and doggy smells than they are to the other individual qualities of Pharaoh.

I will call these individual qualities *tropes*, although tropes, or trope-like entities have been known by a variety of different names, including moments (Husserl 1900), modes (Lowe 1998, 2006; Heil 2003), individual properties (Honderich 1988), ways of being, event-aspects (Achinstein 1983), individual accidents (Simons 1994, following Aristotle) and abstract particulars (Campbell 1981, 1990; and sometimes Williams 1953).

3.1 What is abstract about abstract particulars?

The term 'abstract particular', which Campbell favours, requires some explanation. It is fairly obvious that 'particular' in this context means an entity which is an unrepeatable individual, but the term 'abstract' is less clear, since it must be understood as being contrasted with 'concrete', and not as 'abstract' in the sense that transcendent universals are abstract entities which exist outside space-time. According to Campbell (1981: 126), 'abstract' entities, such as the blueness of painting IKB79, are so called because attending to a trope requires an act of abstraction to set it before the mind; we attend to a trope 'by concentrating attention on some, but not all, of what is presented'. I must focus upon the blueness of the painting, rather than its rectangular shape, or its surface texture in order to apprehend the particular blue-trope. This characterization of the distinction between abstract and concrete in terms of perception or attention is contested by other trope theorists, however. First, because some of them treat tropes as the direct objects of perception: I am directly aware of the particular blueness, the rectangular shape, the particular texture and so on, and so no mental act of abstracting these qualities from the others is required. Early trope theorists such as Husserl (1900) and Stout (1923) postulated tropes as *phenomenological* entities, entities which exist as part of the ontology of experience rather than as the furniture of the mind-independent objective world, and thus they are unproblematically associated with our direct perceptual experience. Furthermore, despite tropes being re-conceived as a category of entities which exist independently of human minds and experience, some trope theorists still maintain that they are the direct objects of perception and this claim sometimes forms the basis of arguments for the existence of tropes (3.4.5). Thus, Campbell's characterization of 'abstract' does not suit everyone.

A better way to understand the 'abstract' of 'abstract particular' is presented by Williams. He rejects 'the many meanings of abstract which … suggest that an *abstractum* is the product of some magical feat of mind, or the denizen of some remote immaterial eternity' (1953: 121) in favour of what he takes to be the "true" meaning (the scare-quotes are his), as something '*partial, incomplete,* or *fragmentary,* the trait of what is less than its including whole' (1953: 122). An abstract particular is a qualitative part of a particular object or event, 'a designation applied to a partial aspect or quality considered in isolation from a total object, which is, in contrast, designated concrete' (1953: 122 n.13). Another related characterization of the required distinction between concrete and abstract is based on the assertion that concrete entities particular objects and events, say are capable of *independent* existence, while abstract entities are not. A particular dog is a self-subsistent object, capable of isolated existence, while a blue trope could not possibly exist without other entities. However, even though these latter formulations are intuitively plausible, they are not without detractors. For instance, Campbell (1981: 127-8) points out that these observations do not establish the metaphysical necessity of the claim that concrete particulars are the minimum entities 'apt for being', or capable of independent existence. It is metaphysically possible for 'free-standing' tropes to exist, he claims; there could be a possible world which contained a single charge trope, or a single blue trope, and such possibilities are made more plausible if one thinks that there are problems with the category of concrete particulars, in quantum theory for example.[2]

In the following discussion about the viability of tropes for a theory of properties, the question of whether a trope can exist independently of other tropes, or of other categories of entities such as a substrate in which tropes inhere, need not be resolved. I will also avoid the question of whether tropes are simple entities, or complex ones formed by entities of more fundamental categories, except where this influences the outcome of other ontological issues. For simplicity, I will presume tropes to be simple entities, although that does not rule out the possibility that some tropes are themselves composed from more fundamental tropes; and I will remain neutral about the question of whether tropes, or a single trope, is capable of independent existence, as it has no implications for what is discussed.

3.2 Resemblance between tropes

A pressing question concerns how an ontology of individual qualities is going to serve as an account of qualitative sameness, or similarity, across disparate particulars at all. As it has been explained so far, the ontology of tropes is one of unrelated particular qualities. Without additional ontological explanation, the blue tropes t_1 - t_{11} in each of the eleven paintings in Yves Klein's 1957 Milan exhibition are no more similar to each other than they are to the smell of Pharaoh the dog, the mass of the easternmost column of the Acropolis, the intensity of Barnard's Star and so on. This was not what the theory of properties *qua* tropes set out to achieve, although in some of its earlier incarnations trope theory is explained this way, thus in effect setting up straw man for its critics to ignite. An ontology of tropes without an account of resemblance between them is an ontology of *bare particulars*: because each trope is a distinct individual quality, such entities lack the ontological resources to be either the same as each other, or to be different (on the basis that there is no sense in which entities can be different if there is nothing which makes them the same or similar). An ontology of bare particulars is implausible and fails as a theory of properties. This was Armstrong's main criticism of tropes (1978a) but in later work (1992) he appreciated that his objection no longer applied (if it ever had).

Not much need be added to trope theory to neutralize this objection: what is required is an additional, fundamental ontological mechanism which determines relations of resemblance holding between tropes. Thus, tropes t_1-t_{11} in the International Klein Blue paintings are exactly similar to each other, while they are more similar to other blue tropes than to orange ones, or to the black of Pharaoh the dog.

There are three main proposals to characterize resemblance between tropes:[3]

1 *Standard Theory*: resemblance between tropes is determined by their natures.

2 *Resemblance Trope Nominalism*: the nature of a trope is determined by its resemblance relations to other tropes.

3 *Natural Class Trope Nominalism*: the nature of a trope is determined by membership of natural classes of tropes, and membership of those classes is not determined by resemblance relations between tropes.

3.2.1 *The Standard Theory: Trope resemblance is primitive*

The first option, favoured by Williams (1963: 608), Campbell (1990: 37), Bacon (1995) and Maurin (2002), takes the qualitative nature of a trope to be an intrinsic primitive feature of it; resemblance between tropes is grounded by the natures of the tropes themselves and is not reducible to anything else. One advantage of the Standard Theory is that exact resemblance can be treated as an internal relation: the existence of the relata (that is, the existence of two tropes) is all that is required for the relation of exact resemblance to hold, or not to hold, between them. Given that tropes have the primitive qualitative natures that they do, resemblance relations also hold between them; resemblance supervenes on the tropes it relates and is available at no extra ontological cost.

If resemblance relations are not regarded as internal relations, but treated as additional relation-tropes holding in each case of resemblance, one might be worried about starting a regress of the kind which Russell warned about in the case of resemblance classes of objects (2.1.2). If tropes t_1, t_2 and t_3 exactly resemble each other, then they do so in virtue of exact resemblance relation-tropes r_1, r_2 and r_3 connecting them; but then relation-tropes r_1, r_2 and r_3 exactly resemble each other, in which case there must be yet more relation-tropes connecting them (Küng 1967; Simons 1994; Daly 1997: 148–53). Is this regress vicious? One can argue that if the intrinsic nature of tropes grounds their resemblance relations, the infinitude is benign even if the relation-tropes are genuine additions to the ontology, rather than their being entirely derivative internal relations as was previously supposed. Despite there being infinitely many relation-tropes, they are all ultimately dependent upon the intrinsic qualitative natures of tropes t_1, t_2 and t_3 with which we began, rather than the dependency being the other way around such that t_1, t_2 and t_3 depend upon an infinitude of resemblance relation-tropes to resemble each other. This direction of dependence is sufficient, it is argued, for the infinitude to be benign (Küng 1967; Maurin 2002: 78). However, neither of these responses would be available were the tropes involved not to have unanalysable intrinsic qualitative natures, and so the infinite regress problem will recur for those who prefer the second Resemblance Nominalist account of trope resemblance, in which the qualitative nature of a trope is determined by the resemblance relations it bears to other tropes.

So, why would one not simply opt for the first conception of tropes as individual intrinsic qualities over the other two which may encounter philosophical problems of their own? One reason stems from an objection that it is not coherent to treat tropes as simple entities under the Standard Theory conception of tropes as particular qualities. On the one hand, Maurin has argued that the simplicity of tropes is required in order for trope theory to offer explanatory utility and genuine ontological novelty; that is, in order that tropes do not collapse into, or become ontologically equivalent to, other ontological categories of which we already have a good understanding. While, on the other hand, her critics claim that 'the trope's tripartite nature [as simple, particular, and qualitative] is a mystery comparable perhaps to that of the holy trinity' which cannot be sustained, alongside the modest truth-maker theory which trope theory aims to uphold (Maurin 2005: 135).

The difficulty arises because both the *particularity* and the *nature* of a trope are intrinsic to the trope itself and this appears to require that a simple entity has parts, or at least that it has dual intrinsic aspects; but by definition such simple tropes cannot have parts or distinct intrinsic properties (Hochberg 1988: 2001; Ehring 2011: 177). The objection is usually presented in terms of truth-maker theory which, in its weaker form, maintains the following principle:

(T) If there exists at least one truth-maker T for <p>, then <p> is true.[4]

Consider two sentences:

   a) 't exactly resembles t*'

   b) 't is numerically distinct from t*'

If t and t* are distinct, exactly resembling tropes, then (a) and (b) are both made true by t and t*. But in that case, t and t* cannot be simple individual qualities because they ground, or act as truth-makers for, more than one proposition. Crucially, (a) and (b) are logically independent of each other and it seems plausible to suppose that logically independent sentences require distinct truth-makers (Hochberg 2001: 178 n.4). A pair of tropes can be numerically distinct and yet not exactly resemble each other and (if we take resemblance to be reflexive) a trope can resemble itself while not being numerically distinct from itself; the truth-values of sentences like (a) and (b) relating arbitrary tropes are determined by the particularity of the relata or their respective intrinsic natures. But if tropes are simple, there are no such differentiable aspects: the tropes either exist or they do not.

One could give up on the simplicity of tropes at this point and treat tropes as complex entities made up of simpler ontological constituents, such as a particular instantiating a universal – an entity which Armstrong would call a *state of affairs* – or as a substrate instantiating a universal. Rather than the particular blue of IKB79 being a simple trope, the particular-painting-instantiating-the-universal-blue is a structured complex entity, a trope constituted by more fundamental entities standing in a certain structure. Such particular, structured complex entities have particularity in virtue of the particular or substrate which partially constitutes them, and intrinsic qualitative nature in virtue of the universal which they instantiate. Thus, they could simultaneously act as truth-makers for propositions about the numerical distinctness of tropes and their exact resemblance. But, Maurin argues, this strategy removes a primary motivation for developing trope theory in the first place: tropes would not be an interesting, potentially fundamental category of entities which could serve as the ontological basis for properties (and, trope theorists would argue, for much more besides) but a derivative category formed out of more fundamental entities, such as 'thin' particulars and universals, for which we already have well-developed theories. Calling such entities 'tropes' does not provide any theoretical gains[5] (Maurin 2005: 134–5).

Alternatively, a novel ontology of structured complex entities could be provided by treating tropes as *particular* qualities instantiated in a substrate (Martin 1980; Simons 1994; Heil 2003). But in this case, even if we presuppose an adequate account of what a substrate is, Maurin (2005: 135) points out that each particular quality instantiated in the substrate must itself be simple: it must be particular in order that a substrate-instantiating-a-particular-quality does not collapse into a state of affairs, yielding an ontology like the one rejected in the previous paragraph; it must be qualitative to give a nature to the substrate; and it must be simple to avoid a regress. If substrate S instantiates trope t^* and t^* is not simple, then t^* itself is another complex of something's instantiating a trope t^{**}, which is itself a complex of something's instantiating another trope t^{***} (and so on). Aside from having little idea about what these 'somethings' might be (having very little idea about what the substrate is in the first place), such a regress could only avoid being vicious if the world were infinitely complex, which is a very restrictive requirement on the way the world might be and consequently not an attractive metaphysical assumption. As Maurin remarks, 'simple tropes have sneaked in the back door' when we were trying to make them complex.

But this line of argument merely tells us that tropes have to be simple if they are to be ontologically useful in their own right – either alone, or in conjunction with a substrate – rather than being derivative, complex entities. It does not answer the objection above that simple tropes with primitive intrinsic natures are inconsistent with truth-maker theory: if such tropes cannot be simple, then an alternative conception of tropes will be needed, or trope theory will have to be abandoned. Before accepting one of these conclusions, however, it will be instructive to investigate whether any of the assumptions which generated it are at fault.

Trope theorists are loathe to abandon the weak truth-maker theory characterized by (T), as this motivates a key argument for trope theory (3.4.3). But the other key assumption that logically independent basic propositions must have distinct truth-makers might be open to criticism: perhaps the logical independence of sentences is not sufficient for the ontological independence of their truth-makers. While Mulligan et al. opt to abandon this principle, a stand for which they do not offer an argument (1984: 296), Maurin argues that the assumption fails in a more restricted way (2005: 134). For instance, consider the sentences 'A is older than B' and 'B is younger than A' which are made true by the same 'external fact' and yet are nevertheless logically independent of each other; 'A is older than B' does not entail that 'B is younger than A' without an additional assumption about the formal properties of the relations involved (namely that one is the inverse of the other). The sentences are materially equivalent – they have the same truth-value as each other in all possible situations – without being logically equivalent. Can we say the same about trope-relating sentences 't is numerically distinct from t*' and 't exactly resembles t*'? Maurin urges that the answer to this question is 'yes' – distinct tropes with primitively resembling qualitative natures make both sentences true – but that claim does not seem as clear cut as in the example of 'is older than' and 'is younger than'. In cases where the reference of terms 't' and 't*' are rigidly fixed, Maurin is right: the truth-values of sentences such as (a) and (b) remain the same relative to each other in all possible situations, just as they do with 'A is older than B' and 'B is younger than A'. However, within a possible situation, if we substitute in other terms for 'A' or 'B', the *relative* truth-values of the sentences 'A is older than B' and 'B is younger than A' does not change. For instance, if substituting 'C' for 'A' renders the first sentence false, it will also make the second one false too. But this does not hold in the case of tropes: substitute 't**' for 't*' into (a) and (b) and the truth-value of one sentence can change while the other stays the same. What determines that is whether 't**' refers to the same particular as 't*' (which makes (b) false), or whether t** is qualitatively distinct from t (which makes (a) false). While Maurin is correct that the sentences retain their relative truth values in all possible situations when the references to t and t* are rigidly fixed, what the truth values of (a) and (b) are for any arbitrary trope(s) depends upon both the intrinsic qualitative nature and the particularity of the tropes concerned; we seem to require that tropes are not simple.

There are two responses at this stage: First, one could argue that we have a sufficiently robust conception of tropes as simple particular qualities to deal with the problem. When sentences like (a) and (b) are considered on a case by case basis as holding between specific trope(s), such tropes fix the truth values of both sentences in all possible situations. It is only when we treat (a) and (b) as instances of universal relations – which is begging the question against the trope theorist – and think that we can substitute one trope term for another that the problems start. For a trope theorist, instances (a) and (b) do not have to behave in the same way whatever their relata, because such sentences pick out a different relation-trope when their relata change.

The second option is to prise apart the particularity of the trope from its qualitative nature, such that only one of these features depends upon the trope's intrinsic nature. Characterizing particularity as external would involve giving constitutive identity and individuation criteria for individual tropes, which are neither primitive nor dependent upon the trope's intrinsic nature. This is an option I will explore in 3.3.1. On the other hand, treating the qualitative nature of the tropes as external to individual tropes themselves would involve jettisoning the Standard Theory in favour of an alternative account of trope resemblance.[6] It is to this option which I now turn.

3.2.2 Resemblance classes of tropes

The troublesome conflict between the primitivist account of trope resemblance and truth-maker theory might encourage trope theorists to opt for an alternative account of trope resemblance. On either account, the problem of shared truth-makers – and thus the potential clash with truth-maker theory – does not arise since what makes it true that tropes t and t* resemble each other is different to what makes them numerically distinct, the former being fulfilled by conditions external to the tropes themselves.

Of the two options, Resemblance Class Trope Nominalism – which regards tropes as having their natures fixed by their resemblance relations to other tropes – is the least popular, most probably because it encounters some serious difficulties which the other views appear to be able to avoid. The first is another occurrence of the resemblance regress (3.2.1). Recall that the Standard Theory could dodge the regress either by appealing to the claim that resemblance is an internal relation, holding in virtue of the primitive qualitative natures of its relata, or by maintaining that the infinitude of resemblance relations generated is harmless, since it is entirely ontologically dependent upon the original resemblance between individual tropes. But if resemblance relations between tropes determine the nature of the tropes themselves, this response is not available, since exactly resembling tropes t_1, t_2 and t_3 have no intrinsic qualitative nature which could determine resemblance relations or allow such relations to be treated as internal. Rather, the dependency runs the other way, with resemblance relation-tropes r_1, r_2 and r_3 determining, and hence being ontologically prior to, the qualitative natures of the tropes t_1, t_2 and t_3 which they relate; so resemblance relation-tropes have to be treated as additions to the ontology. Similarly, the resemblance relation-tropes which hold between resemblance relations r_1, r_2 and r_3 are also additional; and so on. The familiar regress is in progress and this time there is no easy way to stop it.[7]

Secondly, Resemblance Class Trope Nominalism runs into difficulties if there is a trope which does not resemble any other other trope. While the Standard Theory allows the existence of a unique particular quality q*, the Resemblance Class Trope theory implies that such a trope has no nature, since it bears no resemblance relations to other tropes. There are two responses to this: first, one could simply deny that there could be such a trope; that is, one could maintain that, although exact resemblance relations may not obtain between q* and the other tropes in the world, inexact ones do obtain and that is enough to imbue q* with the requisite nature. But that would not account for the case where q* was nothing like any other trope. Second, one could adopt modal realism and say that although there are no actual tropes which exactly resemble q*, there are counterparts of q* in other possible worlds which do exactly resemble q* and that q*'s relations with these are enough to fix q*'s nature. The latter response assumes, of course, that q* is not a necessarily unique trope; that is, that there are no other possible tropes which exactly resemble it, as such tropes would still lack a qualitative nature.

A third difficulty, associated with the problem of unique tropes, is the counterintuitive way in which this account ensures that exact resemblance is a reflexive relation: a trope t resembles itself in virtue of its relations with other tropes (because resemblance is both transitive and symmetric) but this seems to be a convoluted way to ensure the reflexivity of resemblance. Surely, the supporters of the Standard Theory would argue, t resembles itself because of what it is, because of its nature? Again, the resemblance class account will run into difficulties with unique tropes such as q*, since q* is related to no other tropes and so cannot resemble itself. But this difficulty for q* is no worse than the aforementioned problem of q*'s lacking a nature entirely according to Resemblance Class Trope Nominalism and may be solved in a similar way. If modal realism can come to the rescue by facilitating the existence of exactly resembling counterparts to q*, then q* will also resemble itself, albeit in virtue of its relations with transworld counterparts.

Alternatively, one could simply stipulate that resemblance relations are primitively reflexive: it is just a fact about resemblance that q* resembles itself. However, if one adopts this solution, it is not clear how much Resemblance Class Trope Nominalism differs from the Standard Theory: the nature of q* is determined by q* itself, in the latter case because of q*'s primitive nature and in the former because q* resembles itself. In both cases, q*'s nature does not depend upon anything more than q*; the existence of a single trope is sufficient for that trope to have the nature it has (contrary to the position in which Resemblance Class Trope Nominalism started out). In addition, on this revised Resemblance Class view, self-resemblance can plausibly be thought of as necessary to fix the nature of a trope, in which case it is not clear why other tropes are needed to fix resemblance even in cases of non-unique cases since all tropes self-resemble (Ehring 2011: 193). As with the Standard View, the nature of a trope is determined by the trope itself.

3.2.3 *Natural classes of tropes*

Does the Natural Class Trope Nominalist account of resemblance between tropes fare any better than its rivals? On this account, the nature of tropes is determined by their belonging to natural classes of tropes; tropes exactly resemble each other in virtue of being co-members in the same class, or the same classes, of tropes (Stout 1921–3; Ehring 2011: 175–202). A blue trope is blue in virtue of its being a member of the natural class of blue tropes (and other classes associated with blueness, such as colour), and it inexactly resembles tropes of other shades of blue because the trope classes of which they are members partially overlap. Unlike on the Standard Theory, the nature of a trope is not identical to it; instead, a trope's nature is defined as the set of all natural classes of which it is a member. Thus, a trope t* is not identical to its nature, but it is 'of a certain nature' and the trope's nature does not constitute trope t. Ehring draws a loose analogy between his conception of the natures of tropes and the 'nature' of a simple point in a co-ordinate system being fixed by its position in the co-ordinate system: 'That a point could have a "nature" in that limited sense is consistent with the simplicity of the point. The nature of a trope is its "position" in the realm of types.' (2011: 189)

Since the similarity between tropes is grounded by their natures, Ehring maintains that the intuition that this should be the case is preserved, just as it is in the Standard Theory, and as it fails to be when the natures of tropes are determined by resemblance relations between them. Moreover, in contrast to the Resemblance Class Trope Nominalist, the Natural Class account can deal with a unique trope q* having a nature, since such a trope will be in a singleton class – a class with one thing in it – and this class can be identified with the nature of the trope. Second, the formal properties of exact resemblance are respected: it is symmetric, transitive and reflexive even in the case of unique tropes without requiring a primitive assumption about reflexivity, or the acceptance of modal realism; while resemblance is non-transitive as required. Third, although properties correspond to natural classes, these natural classes are particulars, rather than universals. Therefore, although similar objects are similar because each has a trope with numerically the same nature because the tropes are each members of a specific class this sharing of class membership does not introduce a universal since the thing that is shared, the class, is a particular (Ehring 2011: 190 n.21).

One might wonder about Ehring's defence of the particularity of classes at this point: even if natural classes are particulars as he maintains, there seems to be a requirement for repeatable natural classes, or the property of *being a natural class,* and it is not immediately clear that this property can be a particular. This difficulty may turn out to be similar to the regress problems associated with all of the accounts of properties we have encountered so far. Since that makes Ehring's account no worse off than the rest, I will not pursue this objection here.

But what makes tropes members of the natural classes they are members of in the first place? The answer to this question, from Ehring at least, is that they just are; it is a primitive fact about tropes which types of tropes they are. Furthermore, one might worry what makes it the case that the natural classes are the types which they are: What makes class C of which t_1 is a member *blueness,* and class D *charge*? Again, this is a primitive fact about the ontology of tropes which cannot be analysed further: there just are such classes which serve to provide an ontology of types. One might be concerned about these primitive assumptions, but the supporter of this variety of trope theory is arguably in no worse a position than the proponents of other accounts of the ontology of properties: if one asks what determines the range of universals that there are, or what makes the universal *blue* blueness rather than redness or charge, the answer is that 'it just is'. Similarly, the supporter of the Standard Theory in which tropes have primitive natures accepts that which natures there are – and thus which resemblance relations hold between them, thereby determining which types of tropes (or properties) there are – is also primitive. There will be more to say about these assumptions about the range of properties which exists in Chapter 9, but for now it seems that the trope theorist is in no greater difficulty than other property theorists.

A second objection one might raise about Natural Classes of tropes is that although this theory solves the resemblance regress – by getting rid of resemblance relations between tropes – it gets the direction of explanation the wrong way around. Tropes do not fall into the same natural classes because they resemble each other (which is what we would intuitively expect) but they resemble each other because they are members of the same natural classes. Moreover, despite Ehring's claim that his account respects the intuition that the natures of tropes determine their similarity, that is only true because of the way he has defined the nature of a trope. The trope's nature is something largely external to it which the trope partially constitutes, not the intrinsic quality which constitutes it as it would be on the Standard Theory. Intuitively speaking, the 'nature' of a trope is not what we would expect it to be when we began investigating tropes and, although this formulation technically respects Armstrong's intuition that nature determines resemblance, it only does so because the account of 'nature' has so radically altered what the nature of a trope is.[8]

3.3 Individuating tropes

In addition to the question of what makes tropes qualitatively the same or different from each other which we explored in the previous section, we may also be interested in how tropes are individuated: What makes tropes *numerically* the same or different? Such a project might be important for someone who favours an account of concrete particulars as being bundles of tropes, since the individuation conditions of concrete particulars might be taken to be dependent upon the individuation of the tropes which constitute them (Lowe 1998: 205-9; 2003: 82–5). There is extra motivation to answer this question for those who take the qualitative nature of tropes to be an intrinsic primitive feature of them, since (as we saw in 3.2.1) they are faced with an objection to their view which says that tropes cannot be simple, particular qualities, but require separate aspects in order to ground truths about their particularity and the qualitative nature respectively. Since the simplicity of tropes is regarded as highly desirable if they are to do the ontological work their supporters would like (Maurin 2005), it would help to have an independent account of their particularity which did not rely upon their internal nature. Once again, there are three principal options, which are detailed on table 3.1.

Table 3.1 The Individuation of Tropes

Object Individuation (OI)	For all tropes t_1 and t_2 which exactly resemble each other, $t_1 = t_2$ *if and only if* t_1 belongs to the same object as t_2, and $t_1 \neq t_2$ *if and only if* t_1 and t_2 belong to distinct objects.
Spatio-Temporal Individuation (SI)	For all tropes t_1 and t_2 which exactly resemble each other, $t_1 = t_2$ *if and only if* t_1 is at zero distance from t_2, and $t_1 \neq t_2$ *if and only if* t_1 and t_2 are at non-zero distance.
Primitivist Individuation (PI)	For all tropes, $t_1 = t_2$ iff $t_1 = t_2$, and $t_1 \neq t_2$ iff $t_1 \neq t_2$; or, what makes tropes the same or different is a primitive feature about them.

The first, object-based account is unavailable to many trope theorists because it requires the prior existence and individuation of particular objects, which they would prefer to characterize as bundles of tropes. An advantage of trope theory, according to its supporters, is that tropes can provide an ontology of concrete particular objects (and also events, should they be required) at no extra ontological cost. To do so, trope theorists postulate an additional relation – often called *compresence* – which binds tropes together into bundles. This one-category ontology is attractively parsimonious, but it rules out the fairly intuitive suggestion that exactly resembling tropes could be individuated by the objects to which they belong because the individuation criterion would be circular: tropes would be individuated by particular objects, which would themselves be derived from (and individuated by) bundles of tropes. There is arguably nothing wrong with circular individuation conditions *per se* as this is an ontological circle, rather than an epistemological one: the conditions are constitutive, they determine what makes tropes (and objects) distinct or the same, but need not provide an epistemic method by which we find out whether particular tropes, or particular objects are distinct. But this tight circle of interdependence reveals little about the ontological structure of the world: particular objects and tropes are conceptually too close for claims of their interdependence to provide a revelation of philosophical interest.[9] Such considerations do not count against the truth of OI; but it would not provide the one-category *trope* ontology which many trope theorists hope for, and it provides virtually no explanation of the individuation or nature of either tropes or objects along the way.

On the other hand, if one does not aspire to a one-category trope ontology, OI is still available. So, trope theorists who maintain that tropes must be instantiated in a substrate to form concrete particulars could maintain that the individuation of the tropes depends upon the individuation of the substrate (Martin 1980; Heil 2003). However, this suggestion raises the question of whether there are necessary and sufficient conditions for individuating substrata, a problem made peculiarly difficult since these are entities which lack properties by definition; they are true bare particulars. Alternatively, one might suggest that trope bundles, which play the ontological role of concrete particulars, have individuation conditions independently of the tropes which constitute them in terms of the spatio-temporal individuation conditions of particular events (Schmidt 2005). If this suggestion is viable, the individuation of tropes would depend upon the individuation of the bundles which they compose or the events in which they participate. The primary objection to this option from many trope theorists would again be that even if it works, it makes the individuation of tropes parasitic upon the existence and individuation of other categories of entities, thereby ruling out one of the main attractions of tropes that they can characterize the fundamental ontology on their own.

3.3.1 *Spatio-temporal individuation*

If object-based individuation will not do the job, one might try individuating tropes in terms of their spatio-temporal location (SI): numerically distinct, exactly resembling tropes cannot be co-located and numerically identical tropes cannot exist at a spatio-temporal distance from each other. This option has the benefit of being intuitive and not obviously dependent upon the individuation of some other category of entities for which individuation conditions are not yet obvious. (I will presuppose that spatio-temporal location is not especially metaphysically problematic, although one could contest that.) It also has the advantage of providing an independent account of the *particularity* of tropes, which would side-step the objection that tropes cannot act as truth-makers for statements about their particularity and their resemblance relations to other tropes simultaneously. Nevertheless, SI is not a particularly popular option among trope theorists with Jonathan Schaffer, in his earlier work, being its lone proponent (2001).

SI prompts two immediate worries: the first is whether it requires an absolute, or substantival conception of space-time in which tropes are determinately located; the second concerns whether it allows for the existence of non-spatio-temporal tropes and, if it does not, whether that renders it implausible. Schaffer rejects the first charge, arguing that tropes can be individuated spatio-temporally even if the spatio-temporal distances required are regarded as essentially relational. Distance relations, or spatio-temporal location on a non-relational account, would have to be treated as fundamental features of the ontological theory, just as compresence relations are treated as fundamental in order to form bundles of tropes, or resemblance relations are presupposed by resemblance class trope theorists to determine qualitative likeness between tropes. There is no requirement for the existence of substantival space-time in which tropes are located, although such a phenomenon is not inconsistent with the theory. What would not work, or would make the SI account collapse into a variant of OI, is if spatio-temporal relations are *reduced* to relations between particular objects or events. But the trope theorist need not be committed to reducing space-time to matter and change.

The strategies used to answer the second point about the possibility of non-spatio-temporally located tropes might be familiar from the analogous objection to immanent universals in 2.3. Obviously, according to SI, such tropes lack individuation conditions, so the supporter of SI has two options. First, he can endorse a thoroughgoing, spatio-temporally restricted naturalism and insist that the only entities which exist – and could possibly exist – are spatio-temporal ones. Campbell, a former supporter of SI, dismisses this defence for being 'impossibly weak' (1990: 54), although one could respond that there are strong independent motivations to endorse spatio-temporally restricted naturalism aside from preserving a particular account of trope individuation; thus maintaining the naturalist account without this strategy being *ad hoc*. Campbell (1981: 136-7) then proposes a second, 'less drastic' strategy to sustain SI: if non-spatial particulars exist, it is plausible to think that there must be some analogue of the spatial order, and that such an ordering could form the basis of their individuation conditions.[10] However, he later rejected his own suggestion in favour of PI, on the basis that such an ordering would be extrinsic to the tropes and thus is unsuitable to individuate tropes from each other:

> But the orders angels might fall into, of the intensity of their powers [etc] ... while formally they might do the job of providing each with a unique and particularising 'location', seems somehow too extrinsic to carry conviction as to what sets each angel apart from the others. (1990: 56)

Do we care as much as Campbell about having individuation conditions which can deal with angels and their ilk? Many naturalists would say not, among them David Lewis (1986: 73), and Jonathan Schaffer, who retorts: 'Here my naturalistic scruples are pushed to the limit and I must confess to not caring about whether the number of angels dancing on non-spatio-temporal pins has been counted over-formally or not' (2001: 252). In other words, metaphysics does not have to treat everything we can conceive of as if it were possible and account for it accordingly. But despite this bravado, the naturalists might eventually run into problems with numbers, or the properties of numbers, or with trope analogues of alien, or uninstantiated universals, none of which are obviously spatio-temporally located. Analogously to the case of immanent universals, the supporter of SI has the option of accepting modal realism to facilitate the existence of alien tropes, should it be impossible to explain them away, but if there are abstract objects such as numbers, they will not yield to this treatment; whether spatio-temporal individuation conditions can explain these, or explain them away, will be the ultimate test of this method of individuating tropes.

If SI fails, or is thought to be unsuitable, then the trope theorist is left with primitive individuation: tropes just are numerically similar or distinct as a matter of brute fact. I will now argue that neither method is entirely satisfactory without the other.

3.3.2 *Swapping and sliding, piling and scattering*

There are some further points of contention between the spatio-temporal (SI) and primitivist (PI) conditions for trope individuation concerning deeper objections which have also been raised against trope theory in general. In simplified terms, these are that trope theory fails to rule out two 'empty possibilities' (that is, they are *not* genuine possibilities) known as *swapping* and *piling*, situations which would make no causal nor empirical difference to the ontology but are nevertheless allowed. The supporter of SI claims that his account of trope individuation introduces the requisite restrictions, while the supporter of PI objects, either because the restrictions introduced are too strong, or because they do not actually rely upon SI to do the work, and invoke an ontological mechanism which is available to the supporters of PI too.

Swapping

Swapping involves either of two *counterfactual* situations: positional swapping such that a certain blue trope t could have been located in the actual position of an exactly similar (but numerically distinct) blue trope t* and vice versa; and object swapping, that the blue trope in one object (Klein painting IKB234, say) could have been located in another object (IKB79, say) and vice versa. Neither of these swap situations could make any difference causally or empirically – the swapped blue tropes exactly resemble each other – and so it would be impossible to tell the difference between the situation where a swap had occurred and one where it had not. Swapping should be an 'empty possibility' (ie. *not* a possibility) but the ontology of tropes does not disallow swapping as it stands and its critics argue, it should do so to be plausible (Armstrong 1989: 131–2).

One might think that SI has the upper hand here: surely if tropes are individuated by location, this will exclude one trope from swapping places with another exactly similar one? However, this assessment is premature, since swapping involves individual tropes being in alternate positions in different *possible* situations, in distinct possible worlds (if you prefer), while SI only holds within a possible world. To put the point another way, the swapping situation is *inter*-worldly, while the spatio-temporal criterion is *intra*-world. Nor would we want to extend the reach of SI so it held across possible worlds rather than just within them, because to do so would result in our precluding a phenomenon which we do want to allow, known as *sliding*: blue trope t* on IKB79 could have been in a different place, had the gallery owner hung the pictures differently, or had Yves Klein spilt the paint on the floor of his studio, or had the picture fallen onto the floor of the gallery because of an earthquake and so on. Tropes could be in different places from the ones they are actually in and we do not want to have to say that they are *different* tropes as a result; so extending SI to hold across possible worlds would introduce too strong a condition on trope individuation.[11]

Schaffer contends that the best way to rule out swapping is to adopt a counterpart theory of tropes in conjunction with SI: rather than tropes in distinct possible worlds being identical, *counterparts* of actual tropes exist in other possible situations; exactly resembling tropes in other possible situations or worlds are not *identical* to an actual trope (t_1, say), but are *counterparts* of t_1. The counterpart tropes are closer to t_1 in terms of their intra-world spatio-temporal relations than any other trope in their respective possible worlds. With this modal ontology in place, it is not possible that the blue tropes of two Yves Klein pictures could swap, for example: the actual blueness t of IKB79 and the exactly resembling blueness t* of IKB234 could not have been swapped with each other, because the nearest relative or counterpart of t, the blueness located *here* in IKB79 at our world would be the blueness still *here* 'post-swap'. (Those who believe in swapping would want to call the trope in IKB79 in the counterfactual situation t* (or t*'s counterpart), while Schaffer maintains that it is t's counterpart.) The blue trope which would be here in IKB79 is a better counterpart for t, the one which actually is here, than a blue trope which would be *there*, since the former has all the same inter- and intra-world resemblance relations and the same distance relations, as the blueness t which actually is here. Thus swapping is ruled out while sliding is permitted, since a counterpart t' of this blue trope t* could be in a different spatio-temporal position if another factor in the possible world in which t' is located is slightly different from the actual one (Schaffer 2001: 253).

If this strategy works to explain swapping and sliding, it would count in favour of SI. However, Ehring raises two objections. First, he claims, it is not SI which is doing the work but the counterpart theory of tropes and that PI can also invoke counterpart theory to insist that the swapping case is not a genuine possibility. One might respond that it would be more plausible for the supporter of PI to accept transworld *identity* between tropes, rather than invoking additional primitive counterpart relations which could restrict or permit possible situations by fiat (and thus seem ad hoc), or else by borrowing the spatio-temporal criteria to make the counterpart relation work (Ehring 2011: 81, esp. n.17). In fact, one might think that PI is committed to transworld identity, but I will not pursue this line of argument here, since Ehring has a second, more damaging objection. He argues that some cases of trope sliding – which is permitted in the theory – turn into cases of trope swapping and so disallowing these (as SI does) would be *ad hoc*. Ehring's problem cases involve exactly resembling tropes sliding into each other's positions, a change which amounts to a case of swapping. Since sliding is allowed, blue trope t_1 (which is at location l_1 in IKB79) could have been at location l_2 (also in IKB79) had the picture been hung the other way up in the gallery (say); but then it could have been the case that t_1 (which is at location l_1) was at at location l_2 *and* t_2 (which is at location l_2) was at l_1, the tropes t_1 and t_2 could have swapped places. It is worth noting here that the swapping involved in such examples is not swapping of tropes between distinct objects, but swapping between positions. However, this distinction will not make a difference to trope theorists who prefer to treat particular objects as being derivative entities constituted by bundles of tropes, since on this view trope swapping between objects will occur if positional swapping can. Sliding leads to swapping and yet the former is permitted and the latter is not under the spatio-temporal criterion of individuation.[12]

Can the supporter of SI amend the characterization of the counterpart relation to permit the sliding-becomes-swapping cases and reject the problematic swaps? Doing so would involve introducing contextual restrictions, in addition to the criteria of spatio-temporal location and resemblance relations, in order to distinguish when swaps occur and when we have a limiting case of sliding. However, this move amounts to reintroducing a primitive qualitative element into the individuation of tropes; SI requires PI after all.

Piling

A second advantage claimed for SI is that it can rule out *piling*, another undesirable scenario which trope theory has hitherto permitted, that the blueness of the painting (say) is not due to an individual trope, but to two or more exactly resembling tropes which are located, or piled, at the same spatio-temporal location (Armstrong 1978a: 86). The primitivist account of individuation has to rule out such piling scenarios with an additional primitive assumption – to the effect that exactly resembling tropes *do not* pile up at any spatio-temporal locations – but SI already excludes such cases.

As was seen above in the case of swapping however, the supporters of PI are apt to turn such difficulties to their own advantage, since there are some varieties of trope piling we might want to permit. We can distinguish two varieties of piling, *stacking* and *pyramiding*, such that the former is problematic and the latter is not. Stacking involves the presence of exactly similar tropes at the same spatio-temporal location – ten blue tropes (say) when one blue trope would be empirically indistinguishable (there is no difference in intensity, nor hue, for example – while the second, pyramiding cases, involve an increase in 'intensity' as tropes are added. There could, for instance, be a two objects o_1 and o_2 such that o_1 has one mass trope and o_2 has ten piled mass tropes, but where o_2 is more massive than o_1 (o_2 is

10 g and o_1 is 1 g, for instance); the pile of tropes in o_2's case increases the massiveness of object o_2 and as such is a trope pyramid. SI rules out both of these possibilities, but one might argue that we should allow pyramiding as a genuine possibility and reject SI.

SI could be defended on the basis of two points. First, that permitting trope pyramids creates difficulties for predication. We do not want to say that the 10 g object o_2 in the example of the previous paragraph 'has mass 1 g', just because it is composed of a pyramid of 1 g tropes. Second, the trope theorist does not require trope pyramiding anyway, since the pyramid situations can be treated as cases of inexactly resembling tropes instead: in the case of the objects o_1 and o_2, there is simply a 1 g mass trope and a 10 g mass trope and so a trope pyramid is not required. Insofar as piling is concerned, SI seems to be the better account.

Scattering

The third point in favour of SI is that it rules out spatio-temporally *scattered* individuals: for example, it is not clear why PI makes the exactly resembling blue tropes in every International Klein Blue painting numerically different tropes, rather than being one spatio-temporally scattered blue particular. Furthermore, if a trope can be spread out in this way over multiple different spatio-temporal locations, the primitivist theory looks as if it might be in danger of collapsing into something similar to the theory of immanent universals (Schaffer 2001: 249). The crucial difference between trope theory and universals theory would simply be that the blue trope is one simple entity scattered across space-time, while the immanent universal would be wholly present in each spatio-temporal locality in which it was exemplified; that is, in this case, wholly present in each patch of blue. The problem is magnified for proponents of trope theory who base their distinction between tropes and universals purely on the ground that the latter can be multiply located, while the former cannot; now, it seems that if scattering is permitted by PI, then the distinction between tropes and universals breaks down.

It might seem that such scattered individual tropes are odd or too obscure to admit into the ontology, but the case would be analogous to scattered concrete particulars which we do admit: the 2012 Olympic Games, for instance, or the Alliance of Small Island States, are both spatio-temporally scattered concrete particulars. Ehring simply accepts such scattered or 'repeatable' tropes (2011: 91), made plausible on his theory since he rejects the universal-trope distinction which is founded on the multiple-locatability of the former.[13] Alternatively, the supporter of PI can rule out the possibility of spatio-temporally scattered tropes by fiat, with a primitive assumption about the nature of tropes to effect that they are spatio-temporally continuous (which amounts to a concession to SI). On the other hand, SI fares better, it is argued, since tropes cannot be at a distance from themselves and so scattered tropes are not permitted.

However, one might worry that SI allows spatio-temporally scattered tropes too. If tropes are not spatio-temporal points, but can have spatio-temporal extension – that is, they can be spread out in space or time – then a trope such as the blue of IKB234 would be 'at a distance from itself' in some sense.[14] By definition tropes do not have parts, but one outer boundary of IKB234's blue is 77.5 cm away from another boundary. Now imagine 'two' exactly resembling blue tropes located at 50 cm (say) from each other with outer boundaries 77.5 cm apart: could this be one trope on SI? It seems plausible that it could be, since SI only requires that an exactly resembling trope not be at a spatio-temporal distance from itself and it would be begging the question against the scattered blue trope to say that it is. The SI supporter must insist that tropes are spatio-temporal points, or he needs the additional requirement that a trope must be a spatio-temporally continuous quality in order to distinguish the harmless case of spatio-temporal extension from the scattering case. But this latter addition of a qualitative component requires something akin to a primitivist account of trope individuation in terms of a trope's intrinsic nature. SI and PI seem to require each other after all, since PI does not rule out scattering either, or else trope theorists need to rule scattering out from their favoured conception of tropes with a primitive assumption.

3.3.3 Trope individuation: Some conclusions

The question of whether SI or PI is best to individuate tropes has not been resolved, but since this is not a book solely about the formulation of trope theory, I will end the debate here. What does seem clear is that the objections concerning swapping, piling and scattering which had been taken to count against trope theories in general, are not crucially damaging to trope theory itself. However, the way in which a trope theorist chooses to respond to these problems will influence which account of individuation she finds most amenable, whether that be SI or PI, or a combination of the two.

SI has the advantage of offering an account of the particularity of tropes which is external to the trope's qualitative nature and so we can maintain the simplicity of tropes without running into problems with the truth-maker principle (3.2.1). Moreover, SI bans exactly similar tropes from being co-located and so empirically undetectable piles of tropes are not permitted. But, if one takes Ehring's examples seriously, in which swapping can be a limiting case of sliding, there may be a genuine difficulty for SI brought about by its restriction on swapping. Perhaps the supporter of SI could find a suitable response to this complaint, or perhaps they could admit the myriad empty possibilities of swapping as some supporters of PI do, including Ehring himself[15] (2011: 85).

On the other hand, PI requires several additional primitive assumptions to avoid the undesirable phenomena discussed in this section: tropes do not stack, although they might pyramid, tropes do not swap, although they can slide, individual tropes do not scatter themselves out in different regions of space-time, and yet we still do not have an account of when one trope is numerically the same as, or distinct from, another as that too is primitive. One might think at this point that PI would have little to recommend it, at least for those of a broadly naturalist persuasion who presuppose that all the tropes with which their theory deals will be in space-time, or are ordered in an analogous way. However, PI is currently far more popular than SI among trope theorists themselves; not that metaphysical truth can be decided in either democratic or authoritative terms, but the disparity is worthy of note (Campbell 1990; Maurin 2002; Keinänen and Hakkarainen 2014; Ehring 2011).

3.4 Arguments for trope theory

So far, this chapter has concentrated on formulating trope theory and stayed conspicuously silent about arguments in favour of adopting an ontology of tropes, rather than another account of properties. There are two related reasons for this delay: the first is that there are few, if any, compelling arguments in favour of trope theory as opposed to another ontological account of properties; the second is that some recent proponents of trope theory accept this conclusion. This should not be seen as a deficiency of trope theory, however. As a matter of historical accident, trope theorists arrived quite late to the contemporary fundamental ontology debate and by the time the serious formulation of objective trope theories began, the weaknesses and circularity in many arguments for alternative views of properties, such as universals or resemblance classes, had already been exposed. For the most part, the trope theorists are not enthusiastic about making similar mistakes and so they stick to the weaker claim that trope theory is more plausible than its rivals.

3.4.1 The argument from mystery minimization

In light of the difficulties associated with the location of immanent universals, in particular, with understanding how immanent universals can be wholly present in multiple spatio-temporal locations, trope theorists highlight their intuitively simpler idea that distinct qualities exist at *each* location, related by resemblance relations. Related to this is Lowe's argument against the coherence of the multiple location of immanent universals (see 2.2, 2.3).

3.4.2 *The argument for the one-category ontology*

Some trope theorists argue that the category of tropes, along with the relations of resemblance and compresence (and perhaps also temporal priority), is the only one we need for a complete account of fundamental ontology. Trope theory solves the one-over-many problem via trope resemblance and accounts for the existence of concrete particulars in terms of their being bundles of tropes by using compresence. This ontology is attractive on the basis of Ockham's Razor – that entities must not be multiplied beyond necessity – as the one category ontology, if it is plausible, is both more simple and parsimonious than its rivals (Campbell 1990: 17). Nevertheless, unless simplicity or parsimony can be treated as being objective features of the world, the comparative simplicity of trope theory compared to other ontological theories does not guarantee its true.

3.4.3 *The truth-maker argument*

Tropes can act as truth-makers for the truths in our language (Maurin 2002, 2005, 2010a, 2010b). This claim is related to 3.4.2, since it requires that the trope ontology is adequate to play all the ontological roles required to make our sentences true. However, as Maurin happily admits, this does not rule out other ontological accounts of the world from doing the job just as well.

3.4.4 *The argument from the existence of complex concrete particulars*

Maurin (2011) argues that if concrete particulars are ontologically complex – that is, if they are constituted by entities of another ontological category or categories – then the existence of such concrete particulars, such as a particular red ball, requires the existence of at least some tropes. In particular, compresence relation-tropes are required to bind the qualitative features of the ball into a particular, while both avoiding the familiar Bradleyan regress associated with such binding relations and respecting the intuition that the constituents of the particular ball might have existed and the ball not have existed. To avoid the regress, she suggests that the compresence relation must be *semi-external*: its relata can exist without it, but it cannot exist without its relata (2011: 74). That way, the relation does not add to the ontology in such a way as to require further relation-tropes to bind it and it is conceivable that the concrete particulars components could have existed without the concrete particular existing. Criticism of such relations has come from Hochberg (2004). Maurin admits that we could save our modal intuitions by accepting a counterpart theory of properties – the constituents of the particular red ball could not have existed without the ball existing, although counterparts of them could have existed – but she argues that this account also makes more sense if one accepts an ontology of tropes (2011: 76–8).

3.4.5 The perceptual argument

Williams argues that we are directly aware of objectively existing tropes in our perceptual experience (1953: 123), rather than perceiving either universals or concrete particulars. This claim is extremely vulnerable to criticism from alternative accounts of perception and perceptual objects and to the worry that it does not allow for a robust enough understanding of the distinction between appearance and reality (simply because we are aware of the world in terms of tropes, that does not make it the case that there are tropes) (also see Mulligan, et al. 1984).

3.4.6 The inexact resemblance argument

While the theory of immanent universals has difficulty providing an account of inexact similarity between entities, trope theory has no such difficulties because the three accounts of resemblance between tropes can all account for particularity of individual qualities; not all resemblance relations need be exact. This capability alone might be regarded as an advantage for trope theory, although the universals theorist (and other opponents of trope theory) might complain that the advantage is founded upon presupposing the existence of primitive, unanalysable relations, or natures, or class membership. But, in fact, the trope theorist could go further if she wanted and allow possible worlds in which there are no objectively existing relations of *exact* resemblance between spatio-temporally distinct entities. In such worlds, while particular qualities do resemble each other, qualitative similarity is never exact; one might regard the concept of exact similarity as an idealized limit which never actually obtains. While this rejects the intuitions of the one over many argument, it does facilitate enough inexact similarity in the mind-independent world to account for the apparent truth of the 'Moorean fact' of qualitative sameness between spatio-temporally distinct particulars. It also permits accounts of causation in which instances of cause and effect can be treated as singular occurrences, rather than essentially being instances of causal laws. The compatibility of trope theory with a plausible account of nature which is causal while not being law-governed (in addition to its being able to give an account of objective laws should exact resemblance relations hold) might be regarded as an advantage for trope theory over that of universals; trope theory permits a greater range of ways in which the actual world might turn out to be. These issues will be discussed in greater detail in Chapter 8.

3.4.7 Arguments from the utility of tropes

Tropes are regarded as the most plausible candidates to act as causes and effects and may also solve the problem of mental causation. Enduring tropes can, it is argued, provide a physical link between cause and effect which is more plausible than the nomological or spatio-temporal connections postulated in other theories (Ehring 1997). Mental tropes may be able to retain their causal efficacy and genuinely cause physical behaviour, despite the physical being causally closed, because their particularity may preclude their being identified with, or reduced to, physical tropes (Ehring 2011, Chapter 5; MacDonald 1989).

A thorough evaluation of these arguments would take us outside the scope of this book and so I will let these claims stand, rather like advertisements for trope theory. Just like advertisements, be warned that the claims made may not stand up to scrutiny, although I will leave it for further reading to find out why.

In general, the main difficulty for trope theorists, at least according to their rivals, is the number and variety of primitive assumptions which the trope theorist requires to get her theory to work as an adequate ontology of properties. These, as we have seen, may vary somewhat according to which formulation of trope theory one prefers, but the extent of them is always notable when compared with property theories based on universals or classes.

Perhaps, though, the trope theorist might respond to this objection that she is simply being more honest: the claims which trope theorists have explicitly made to ensure resemblance relations (say) have simply been 'built into' the ontology of universals; universals are just the type of entity required to play the property role because we have postulated the existence of exactly the kind of entity which we need. Moreover in doing so, the trope theorist might point out, the universals theorist has presumed the existence of repeatable entities which have counterintuitive spatio-temporal properties and an opaque relationship with the particulars which instantiate them. The universals theorist's tactics are no less presumptive than the trope theorist's policy of making her assumptions clear at the start.

FURTHER READING

In favour of tropes:
Campbell 1981; Simons 1994; Bacon 1995; Maurin 2002, 2010; Ehring 2011.
Some objections and discussion:
Armstrong 1992; Daly 1997; Lowe 2006; Benovsky 2014.

Suggested Questions
1   What is the best account of resemblance between tropes?
2   How much should trope theorists accept as primitive fact? Does the fact that they do so put trope theory at a disadvantage in relation to other accounts of properties?
3   What makes one trope numerically the same or distinct from another?
4   Why does the primitivist account of trope resemblance clash with the truth-maker principle? Can this difficulty be satisfactorily resolved?
5   Why does the resemblance class account of trope resemblance lead to a regress and the primitivist theory avoid it?
6   Is there a good argument for the existence of tropes?

Notes

1   There is some debate about the existence of ordinary middle-sized tropes such as these, but I will not cover the debate here. See Campbell (1990: 136–9); Ehring (2011: 91–7).
2   For instance, see Ladyman, et al. 2007 and Kuhlmann 1999, 2010 for a discussion of concrete particulars in quantum theory.
3   I will follow Ehring's (2011, Chapter 6) terminology.
4   Here <p> stands for whichever entities turn out to be bearers of truth value, such as propositions, or sentences. A more maximal, and consequently less plausible, principle is proposed by Armstrong (2004: 5): <p> is true *if and only if* there exists at least one truth-maker T for <p>.
5   In opposition to this view, Cumpa (2012) argues that states-of-affairs can be a fundamental category despite having constituents. To some extent, the difference of opinion between Cumpa and Maurin turns upon what it means for a category of entities to be 'fundamental', which is not a discussion which I have space to enter into here.
6   Maurin (2002, Chapter 5; 2005: 138) rejects this suggestion.

7 I will have more to say about resemblance nominalism in Chapter 4.

8 It seems open to the Resemblance Class account to maintain that the *nature* of a trope consists in its resemblance relations to other tropes, to avoid Ehring's criticism. This characterization is fairly implausible as an account of nature, but perhaps not more so than Ehring's characterization of the nature of tropes as natural classes.

9 There are other purported cases of ontological interdependence where the phenomena related are seemingly very different, such as space and time; or matter and space; or time and change and so on. Another, more interesting case of ontological circularity between properties and laws or causation will be discussed in Chapters 8 and 9.

10 Note, that while I have been talking in terms of spatio-temporal conditions, Campbell talks in terms of spatial conditions. Nothing turns on the difference in the present discussion, although, obviously, the existence of purely temporally located entities would be counterexamples to the spatial view but not to the spatio-temporal view.

11 This solution also presupposes a questionable account of space-time which applies across possible worlds.

12 Ehring (2011: 81–5) sets up the example with a gradual progression of cases in which the tropes approach each other's positions, each of which is allowed until they actually swap places with each other. Unfortunately, space does not permit a full examination of his examples here.

13 See Giberson (MS) against this position.

14 The question of whether simple particulars can be extended is a controversial one which I will not consider in detail here. See Hudson 2007, McDaniel 2007.

15 Ehring maintains that swapping does not put trope theory at a disadvantage with respect to some other accounts of properties: Armstrong, for instance, permits that thin particulars (the entities which instantiate universals) can swap (Ehring 2011: 86; Armstrong 1997: 107–8).

4

Properties as sets or resemblance classes

The plausibility of properties as sets or classes of concrete particulars is discussed, focusing on class and resemblance nominalism. First, the concept of nominalism is clarified and variants of it described. I discuss whether modal realism is sufficient to alleviate the Coextension Problem, or whether a more finely-grained ontology of properties is required which threatens to outrun existing particulars, even on a modal realist view. Goodman's problems of Companionship and Imperfect Community are raised for resemblance nominalism and I investigate Rodriguez-Pereyra's and Lewis's solutions. I then discuss Lewis's class nominalist theory of properties and his primitive assumption about the existence of perfectly natural properties. I conclude that although both versions of nominalism are complicated, they become so by broaching questions which universals theory and trope theory do not touch, and so should not be considered worse off as a result.

While the previous two chapters have been concerned with theories which postulate the existence of specific entities to play the role of properties – universals and tropes – in this chapter I will consider a group of property theories which try to avoid this ontological commitment. An object's *being blue* (say) is explicable in terms of the object's belonging to the set or class of blue things; while an object's *having a mass of 1 kg* is in virtue of its being a member of the class of massive particulars and more specifically, being a member of the subset of that class, the set of particulars with a mass of 1 kg. There are concrete particular individuals – particular dogs, doors, trees, tortoises, atoms, planets, pictures and perhaps also particular events such as your next birthday (whichever one that is), the death of the last Neanderthal, Yves Klein's 1957 Milan exhibition and the formation of the Moon – and these particulars are members of classes or sets in virtue of which they count as types of particulars. There are several ontological accounts of similarity based upon this general idea. In some, properties are *identified* with classes of individuals, and in others they are not so that there are classes, but no properties as such. Also, some theories attempt to give an account of *why* particulars are members of the classes which they are, in terms of resemblance relations holding between the particulars – these, unsurprisingly, are the *resemblance class* theories – while others treat class membership as a matter of brute fact. But these accounts are unified by there being no additional entities corresponding to properties or qualities over and above the apparatus of concrete particulars and the sets or classes of which they are members.

4.1 Why not 'nominalism'?

These set- or class-based accounts of properties are frequently grouped together under the name '*nominalism*', but there are good reasons not to use this term when more specific names for the different theories are available. The primary reason for this is that 'nominalism' is used for two distinct species of ontological denial: first, by those who deny that universals exist, which I will call nominalism$_1$; and second by those who deny that abstract objects exist (nominalism$_2$). Secondly, once this distinction between nominalism$_1$ and nominalism$_2$ is in place, it cross-cuts the different ontological accounts of properties under consideration. Table 4.1 summarizes the positions which different property theorists *can* take.

The account of properties as sets or classes sketched so far is clearly a form of nominalism$_1$, but since sets themselves are often regarded as abstract objects (albeit *particular* abstract objects, unlike transcendent universals), it is not a form of nominalism$_2$.[1] (Note that I have avoided calling such sets 'abstract particulars' in order to avoid confusing them with tropes (3.1).) However, trope theorists also count as nominalists$_1$ since they deny the existence of universals, while the consistency of their theory with nominalism$_2$ will depend upon how similarity between tropes is determined. Furthermore, the supporter of immanent universals might count himself as a nominalist$_2$ because his ontology of universals is spatio-temporally located, while obviously not being the kind of nominalist$_1$ who claims that universals do not exist. The term 'nominalism' used without qualification is unhelpful,

Table 4.1

	Nominalism₁ Denies Universals	Nominalism₂ Denies Abstract Objects
Sets/Classes	Yes	No
Tropes (Class Theories)	Yes	No
Tropes (Standard Theories)	Yes	Neutral
Universals (Transcendent)	No	No
Universals (Immanent)	No	Neutral

and cross-cuts too many philosophical positions to be useful. Although 'nominalism' appears frequently in the literature, I will reserve it for the denial of universals involved in 'nominalism$_1$', and I will avoid the term entirely if I can refer more specifically instead.

4.2 Varieties of class theory

Many theories make class membership the essential feature of the explanation of the qualitative similarity of distinct particulars. But these theories diverge sharply due to the accounts which they give of why particulars belong to the classes that they do, what it is – if anything – that determines class membership. The different varieties of nominalism which give an account of properties in terms of classes or sets of concrete particulars are shown on Table 4.2.

Both concept and predicate nominalism depend upon the existence of thinkers or speakers to classify particulars into sets; thus, properties and the distinctions between them are dependent on language, or meanings, or theories rather than marking divisions in the objectively-existing world which would exist whether or not humans (or other sufficiently sophisticated thinkers) did. While it is plausible to regard some groups of properties as being dependent upon human thought or language, such as those concerning furniture, television programmes, literary criticism and so on, there are many such as water, gold, hydrogen, mass and charge which appear not to be dependent in this way. For those who want a realist account of properties – that is, an account of properties as entities which exist independently of us – the failure of objectivity is a devastating flaw in concept and predicate nominalism. In the case of concept nominalism, this

Table 4.2   Varieties of Nominalism

Name	Description
Concept Nominalism	Particulars belong to a specific class if they satisfy a certain concept.
Predicate Nominalism	Particulars belong to a specific class if a certain predicate applies to them.
Ostrich Nominalism	There is nothing in virtue of which concrete particulars are the types of particulars which they are.
Mereological Nominalism[2]	A property P is identified with the aggregate, mereological sum, or fusion of all the particulars which have P.
Class Nominalism	Particulars have a property in virtue of belonging to a class of particulars.
Resemblance Nominalism	The concrete particulars in a class are members of that class in virtue of resemblance relations holding between the members.
Resemblance Class Trope Nominalism	Tropes are similar to each other in virtue of resemblance relations holding between them (3.2.2).
Natural Class Trope Nominalism	Tropes are similar to each other in virtue of their membership of natural classes (3.2.3).

objectivity might be reinstated by treating concepts as existing objectively, perhaps as abstract entities in a similar manner to transcendent universals; although to be plausible, such a move would require a more detailed account of what sort of entities these objectively existing concepts are. Moreover, if the motivation behind nominalism is to avoid the postulation of universals or universal-like entities, then explicating concepts as objectively existing abstract objects which can be satisfied by different concrete particulars seems to reintroduce entities very similar to universals by the back door; the status of objective concept nominalism as a form of nominalism is in jeopardy.

Furthermore, predicate and concept nominalism are also criticized because the criteria they provide for class membership get the direction of explanation the wrong way around: an object should not belong to the class of blue things *because* it satisfies the concept blue, or because the predicate 'is blue' applies to it, as these theories respectively assert; rather, the concept blue should be satisfied, or the predicate 'is blue' apply to an object *because it is blue* (Armstrong 1978a). Again, this criticism has greatest traction if one thinks that the world determines which types of things there are independently of us, rather than qualitative divisions being, at least in part, a product of our language or thought. What one thinks about the latter claim depends upon how strong the reasons are for reifying properties, or for thinking that there are objective qualitative joints in nature.

There will be reason to return to non-realist alternatives of nominalism in Chapter 9, when we consider how we know about, or can justify beliefs in, which properties there are. However, the current chapter will be restricted to considering the most viable versions of nominalism which treat properties as objective, mind-independent entities – class nominalism and resemblance nominalism – since these are in straightforward competition with the other realist ontological accounts of properties which have already been considered.

4.3 Properties as classes and resemblance classes

Class nominalism identifies properties with sets or classes of entities (I will use 'set' and 'class' interchangeably) such that *blue* just is the class of all blue things, while the property of *being a dog* is identified with the class of all dogs. As with Natural Class Trope Nominalism considered in 3.2.3, there is no further story to tell about why individuals belong to the classes that they do, that is just a primitive fact about the world.

Resemblance nominalism on the other hand accounts for class membership in terms of resemblance relations holding between the members in a class. There are two principal versions of the theory: in the first, class membership is determined by resemblance relations which hold equally between all and only the particulars in the class; on the second, membership is determined by resemblance relations holding between the members of the class and an elite paradigm set of particulars which are also members of the set. Furthermore, the resemblance nominalist can choose whether properties are to be identified with resemblance classes, or whether qualitative similarity between particulars should simply be treated as their belonging to the same class. I will remain neutral between these latter options here.

The criteria for a class being a resemblance class, on the first conception of resemblance nominalism in which resemblance holds equally between the members of a class, were originally specified by Carnap (1928: 70). These can be adapted as follows:

A class of objects C is a resemblance class *if and only if*
(1) each member of class C resembles each other to some degree;
(2) no non-member resembles a member of class C to that degree.
(Goodman 1977: 114–15; Manley 2002: 77)

As it stands, this characterization of resemblance is open to two serious objections which I will consider in 4.5 and will need considerable refinement for this account of properties to be viable.

Alternatively, we might say that a resemblance class contains a paradigm collection of privileged members, sometimes called 'exemplars' or 'standard objects'. For instance, Price suggests that 'the exemplars for the class of red particulars might be a certain tomato, a certain brick and a certain British post box' (1953: 20). Every particular which resembles these exemplars at least as closely as they resemble each other is thereby a member of the class. On the other hand, we could require that each member of the class resembles the particulars in the paradigm more than any non-member resembles all of the paradigm particulars. But what makes a paradigm count as such? We may ask what privileges certain particulars, and on what basis. In some classes, the class of white things for instance, or the class of electrons, it seems that one particular or group of particulars is no better than any other as an exemplar or paradigm class. What counts as a paradigm in such classes, or is the paradigm just stipulated to be any arbitrary subclass of members in order to keep this version of resemblance nominalism alive? Furthermore, what makes a group of particulars count as a paradigm for one property rather than another? If we can characterize resemblance in egalitarian terms as a relation holding between each member of a class equally, it is not obvious that we will require the potentially problematic paradigms at all. In short, until alternative views have failed, it is not clear whether the paradigm version of resemblance nominalism is either viable or well-motivated and so I will not attempt to defend it here (see Rodriguez-Pereyra 2002, Chapter 7).

4.4 The coextension problem

Both resemblance and class nominalism are subject to a serious objection which was for some time considered fatal to these accounts of properties: What if the classes associated with two properties happen to contain exactly the same particulars? According to the identity criteria of classes or sets,

sets are identical if and only if they have the same members, but we do not want to identify properties simply because they happen to be exemplified by exactly the same concrete particulars.

To use a traditional, and rather contentious, example, the class of all living creatures with hearts contains the same particular creatures as the class of all living creatures with kidneys, but one would not want to identify the properties of *being a living creature with kidneys* with *being a living creature with a heart*. The example is contentious for at least two reasons: first, the example is flawed, since the sets are not actually coextensive because flatworms have hearts and not kidneys (although they do have an organ called the nephridium which performs an analogous role to the kidneys); second, one might challenge the coherence of the example on the basis that the properties invoked are *relational*, holding between a living creature and its heart and between a living creature and its kidneys, so as such they are not coextensive. However, whether or not an actual, uncontentious example can be found, the mere possibility of such coextensive properties is sufficient to provide a counterexample to the accounts of properties as sets as they are currently formulated. Imagine, for example, a world in which all the particulars which have a mass of 1 kg are also a specific shade of blue (International Klein Blue, for example) and vice versa; there would be *one* class of particulars for intuitively *two* properties. Divisions between properties are intuitively more fine-grained than divisions between sets and so it seems that properties cannot be defined in terms of the extensions of sets or classes, that is in terms of the particulars which have them. (Similar objections apply to attempts to explicate the meanings of predicates in terms of their extensions.)

There are two hard-headed reactions to these counterexamples. The first denies that such examples are counterexamples and insists that properties *would* be identified in such cases, despite our intuitions to the contrary. No matter how dissimilar *being International Klein Blue* and *having mass of 1 kg* seem to be, there would be no reason to individuate them, if the particulars which had them could never actually be distinguished. We simply have two predicates, 'having mass of 1 kg' and 'being International Klein Blue', which apply to the same entities. This broadly empiricist response maintains that there can be no ontological difference without a difference which is detectable from our epistemic perspective, and for most property theorists, this places too strong an empiricist constraint upon the content of our metaphysical theories to be plausible and leaves the ontology of properties too coarse-grained. The second hard-headed response accepts the coextension counterexamples but maintains that the failure of an extensional set theoretic account of properties to provide identity criteria for properties – that is, the failure of set theory to tell us when properties are the same as each other and when different – should make us suspicious about making an ontological commitment to properties at all. Such a response arises from the view that identity and individuation criteria are essential for commitment to an ontological category of entities, following Quine's slogan, 'no entity without identity' (1969: 23). But, like the empiricist commitment to identifying properties if they are exemplified by the same actual particulars, Quine's condition is also regarded as being too strong a constraint upon our ontology. Moreover, even if one does subscribe to the strong Quinean requirement for identity criteria, it is far from clear at this stage that there are no better identity criteria for properties to be found. The sceptical conclusion is premature until other responses to the problem are considered.

The intuition driving the claim that one class of actual particulars can coincidentally have two distinct properties is a modal one. In the restricted example above, in which all and only particulars which are International Klein Blue also have mass of 1 kg, one is driven to distinguish the properties by the intuition that it is possible for something of IKB colour to have a mass greater or less than 1 kg, or for something with a mass of 1 kg to be a different colour from International Klein Blue. Accordingly, David Lewis (1986: 50–2) proposed that the problem of class coextension in class and resemblance nominalism can be resolved by including possible particulars in the class too and this requires a commitment to modal realism, the metaphysical view that possible worlds, and the entities in them, exist in the same sense as the actual world. A class which includes all possible objects with a mass of 1 kg in addition to the actual objects of 1 kg will not be coextensive with – that is, it will not contain all the same members as – the class of all particulars which are International Klein Blue. The coextension problem is solved by admitting possibilia into the classes in addition to the particulars of the actual world.

There are two primary objections to this solution to the coextension problem. First, one might have general qualms about the plausibility of modal realism, or about the degree of ontological commitment it requires. Modal realism appears to permit the existence of sufficiently many classes to provide a fine-enough-grained ontology of properties to be intuitively plausible, but the price of this is considerable ontological commitment. One might wonder whether a more economical ontology could be found in preference to accepting that possible worlds and the entities which they contain exist in the same sense as actual entities. One advantage which the class and resemblance nominalists had been able to claim for their theories at the outset was the ontological economy of characterizing properties solely in terms of concrete particulars and classes, with some primitive resemblance relations added in the latter case. But, if realism about possibilia is required to make this characterization work, then the avoidance of entities specifically postulated for the explanation of qualities, such as universals and tropes, loses its economical edge. If possible worlds are reified, then even the supporter of the abundant ontology of transcendent universals might be able to plead in favour of the relative parsimony of his own theory, or at least to claim that he is doing just as well. Of course, the supporters of modal realism maintain that their additional ontological commitment is worth it for a host of reasons unconnected to the defence of a nominalist account of properties, but there is insufficient space to evaluate these here (Lewis 1986). What is clear is that, although the additional commitment does nothing to harm the plausibility of the nominalist accounts of properties under discussion (presuming that modal realism itself is tenable), the nominalists who take this route away from the coextension problem can no longer claim to have ontological economy on their side.

The second reservation expressed about the modal realist solution to the coextension problem concerns its sufficiency. Are there distinct properties which are *necessarily* coextensive? Even when classes include possible and actual particulars, there may be intuitively distinct properties exemplified by all and only the members of a single class. Familiar examples of such properties include mathematical or geometric ones, such as *being triangular* and *being trilateral* which would both correspond to, or be identified with, the class of all possible and actual individual triangles. Another necessarily coextensive pair of properties is *being Socrates* and *being a member of the singleton set SOCRATES* (that is, being a member of {Socrates}): in every possible situation in which Socrates exists, his singleton set exists and Socrates is a member of that set, and so we have intuitively distinct properties and yet there is a single class corresponding to them (that containing Socrates and all his counterparts) (see Fine 1994).

How should class and resemblance nominalists respond to these examples? They seem to have two choices if they want to maintain their account of properties based on classes: first, they can deny that examples such as those above are distinct properties; or else they can deny that the properties are necessarily coextensive and thereby deny that more than one property corresponds to, or is to be identified with, a single class. The first strategy of identifying necessarily coextensive properties is much more plausible than it was in the case of the original counterexamples which motivated the coextension problem, in which intuitively distinct properties were found to be coincidentally associated with a single class of *actual* particulars. One can more reasonably claim that there is only one property to be identified with the set of actual and possible triangles, although there are two or more predicates 'is triangular' and 'is trilateral' which apply to the individuals in that set.

Furthermore, as was suggested in the case of 'is a living creature with a heart' and 'is a living creature with kidneys', one might also reject some apparent counterexamples because the predicates apply in virtue of relational properties which are not necessarily coextensive. For instance, 'being trilateral' and 'being triangular' apply in virtue of relations between the sides of a particular and the angles of a particular, respectively, and thus the counterexample does not hold (Rodriguez-Pereyra 2002: 100). If this strategy is successful, then the stock of counterexamples will be considerably reduced.

There is an undesirable implication of accepting this response, however. If one removes the one-to-one correspondence between properties and predicates in order to sustain necessary coextension as the identity criterion for properties, this move will complicate the development of a theory of meaning in terms of which properties – that is, in terms of which classes of actual and possible particulars – there are. There would be much to recommend an account of properties which provided the meanings or semantic values of predicates, but the response currently under consideration makes predicates more complicated than they appear, or associates distinct predicates with the same property. On a 'simple' property-based view of meaning one would think that the meaning of 'is a triangle' would be given by the property of *being a triangle* (that is, the class of all possible and actual triangles). But according to this response, the meaning of 'is a triangle' is either given by a different property to that which gives the meaning to 'is triangular', or it is not provided by the property of *being a triangle* at all, but by *being triangular*, a relation which those particular shapes bear to their vertices. The meanings of some one-place predicates would be given by relations, or else the meanings of some apparently synonymous (or nearly synonymous) predicates would diverge because they would be associated with different classes of particulars.

The second strategy also dismisses the examples of necessarily coextensive properties; not, as in the first response, because they are not distinct properties, but because the problematic pairs in the examples are associated with distinct classes of actual and possible particulars. For example, one might claim that the domain of possible worlds is broad enough to accommodate some logically, or set-theoretically, counterintuitive worlds in which (say) Socrates can exist (or, more accurately, a counterpart S_1 of Socrates exists) and yet the singleton set containing Socrates does not exist. Thus, that counterpart S_1 is not a member of the class identified with *being a member of the singleton set {Socrates}* and so the classes *being Socrates* and *being a member of the singleton set {Socrates}* would not be necessarily coextensive. Such a suggestion might seem outlandish, but it is not without a precedent. For example, the axiom of choice in set theory asserts that the selection of one member from each of a collection of disjoint sets will itself result in a set. It may seem counterintuitive to deny such a claim – select one from each of a collection of sets and you will get a set – but many philosophers of mathematics have been troubled by the axiom of choice and thought it too risky to accept as an assumption. Similarly, claims which are central to classical logic, such as the law of the excluded middle, or the law of non-contradiction, have also been denied. Presumably then, the domain of possibilities accepted by the modal realist would have to make room for such worlds, and those in which (the counterpart of) Socrates does not have a singleton set could also be among them.

This response might run into trouble however, since it will have to find a way to allow logically or set-theoretically counterintuitive worlds alongside the broadly set-theoretic accounts of properties under discussion. (This is an instance of a more general problem with ontological categories and possibility, discussed in Allen 2015.) That may be achievable if its proponents could claim that entities within such worlds could be members of inter-world classes (such as those which determine property identity or resemblance classes) while no intra-world sets existed in certain worlds. Alternatively, one might be able to treat the set-theoretic constructs of the property theory as a form of meta-language not affected by the strange set-theoretical or logical goings on within some possible worlds. If this cannot be done, there is a danger that such a strategy will ultimately be self-defeating, since it could destroy the set-theoretic mechanics of the account of properties which it was proposed to defend.[3]

4.5 More difficulties for resemblance nominalism

In addition to the Coextension Problem, which afflicts both class and resemblance nominalism, resemblance nominalism suffers from two other serious difficulties which were pointed out by Goodman (1977: 116–18). Recall the two conditions by which resemblance nominalists define a resemblance class:

A class of objects C is a resemblance class *if and only if*
(1) each member of the class C resembles each other to some degree;
(2) no non-member of C resembles a member of class C to that degree.

Do these two conditions jointly provide necessary and sufficient conditions for C being a property class: (i) there being a property P which is shared by all the members of class C; and (ii) C being such that all particulars which have P are members of it? If the characterization of resemblance nominalism in terms of (1) and (2) is to work, then this must be the case, but Nelson Goodman showed that (1) and (2) are not necessary, due to the *Companionship Problem*, nor are they sufficient due to the *Problem of Imperfect Community*. I will discuss these problems in turn and consider whether the resemblance nominalist has a plausible response.

4.5.1 The Companionship Problem

Even in his first formulation of resemblance nominalism, Carnap noticed that his account of resemblance classes did not work if certain 'unfavourable circumstances' obtained. If a certain property F occurs only in particulars in which another property G also occurs, then the class of F particulars cannot satisfy both (1) and (2) if it is to exist. The class of F particulars is either part of (a subset of), or coextensive with, the class of G particulars, so (1) is satisfied because all the elements of the F class resemble each other to some degree, but (2) is not satisfied because the members of G (that is, some non-members of F class) resemble members of the F class just as much as members of the F class resemble each other. All the particulars in the F class are also in the G class in virtue of their resembling G particulars to some degree. To use a less formal example, imagine that all red things are also square, but that some square things are not red, so that the class of red particulars is a subset of the class of square particulars. In this case, the red-class will satisfy (1), but it will not satisfy (2) because all the members of the square-class will resemble members of the red-class too. However, it seems intuitively obvious that the red-class should not be barred from counting as a property class in virtue of not satisfying (2), a failure which is simply due to the coincidence of every red particular having a companion property of being square; so the conjunction of (1) and (2) is not necessary for a class to be a property class.

At this point, one might be tempted to respond that this objection is missing an obvious point: the members of the square-class do not resemble the members of the red-class in the correct respect; that is, the former resemble members of the red-class because each one is square, not because they are red, and so if we distinguish the *respect* in virtue of which the particulars resemble each other, the problem will be solved. Unfortunately, this easy response to the Companionship Problem is begging the question. What are the 'respects' which we should distinguish? What entitles us to talk in terms of particulars resembling each other in virtue of being red, or in virtue of being square? In the present context, such distinctions must depend upon which properties a particular has, and that is determined by the resemblance class to which it belongs in virtue of its resemblance relations to other particulars in the class. We cannot sort out problems with intuitively 'overlapping' resemblance relations by appealing to the very entities which the resemblance relations were introduced to explain. Until the ontology of resemblance classes is in place, we cannot distinguish red from square on this account of properties, nor talk of particulars resembling each other *qua* red, or *qua* square.[4] There is no easy solution to the Companionship Problem.

For his part, Carnap chose to simply assume that such overlap between property classes would not occur, acknowledging that if it did his account of what he called 'similarity circles' (or classes) and we are calling 'resemblance classes' would not succeed (1967: 112–13).

> If there are no systematic connections between the distribution of different colors [or other qualities], then this unfavorable case... becomes less likely the smaller the average number of colors [or other qualities] of the thing and the larger the total number of things is. Let us assume that in our case the unfavorable conditions are not fulfilled. (1967: 113)

Even though Carnap's theoretical aim was very different to ours, since he sought to characterize similarities between *experiences* rather than to formulate an account of realistically construed objective properties in terms of resemblance classes, it seems fair to say that his assumption is not justified. As Goodman points out, there is no way from within a system to determine whether companion properties occur or not, and Carnap's assumption also requires that there not be any systematic connections between qualities (1977: 117). But this latter constraint arbitrarily rules out the very cases where the problem is most likely to arise, since systematic connections between qualities – between colours, for example – would make companionship more likely. Similarly, treating both determinable and determinate properties as genuine distinct properties will result in companionship, since (for instance) the class of objects with 1 kg mass will be a subset of the class of all massive particulars; *having mass* will be a companion to *having mass of 1 kg*.[5]

However, one could respond in support of Carnap here that, unlike in the case of objective property companionship, his own project need not answer the Companionship Problem, nor the Problem of Imperfect Community which will be considered in the next section, since both problems require an external viewpoint, an objectively existing distribution of properties over particulars to which the classes of qualities (which are, in Carnap's project, capturing similarities in experiences) could fail to match up. But such an external viewpoint is illegitimate in Carnap's system and so the intuitions which drive Goodman's problems are ill-founded.[6]

It is quite easy to conflate the Companionship Problem with the Coextension Problem which was discussed in 4.4, and there are some cases of this mix-up in philosophical discussions on the problem.[7] The most likely reason for this confusion is that in some cases the Coextension Problem occurs when the

Companionship Problem does; namely, when the companion properties apply to exactly the same particulars and thus when the class of F particulars is coextensive with the class of G particulars. Recall that the Coextension Problem arises because a class associated with one property could contain exactly the same particulars as a class associated with another property; thus we would have intuitively two properties F and G associated with one class, and the class and resemblance nominalists would have no way to distinguish them. But, while the Coextension Problem is concerned with the classes of particulars which are associated with specific properties, the Companionship Problem is concerned with the *structure* of the resemblance relations between particulars, both within a class and with particulars which are not members of the class. As such, only the Coextension Problem presents a difficulty for the class nominalist, since he treats facts about class membership by particulars as primitive.

4.5.2 The problem of imperfect community

Goodman's second objection was that conditions (1) and (2) on resemblance classes are not sufficient for a class to be a property class; that is, they can both be satisfied and yet the individuals in the class do not share a common property. To simplify, let us consider a class I of particulars a, b, and c and properties F, G and H. The property distribution between the particulars is in Table 4.3.

Table 4.3   Imperfect Community

a	b	c
F	–	F
G	G	–
–	H	H

As can be seen from Table 4.3, each particular has two of the three properties and (1) is satisfied because each particular in class I resembles each other particular to some degree. Let us also assume that there are no other particulars with any of the properties F, G, and H (nor do a, b and c have other properties), so that condition (2) is also satisfied, since there are no non-members of the class I which resemble any of members a, b and c.[8] The class I = {a, b, c} satisfies the two conditions to be a resemblance class, but as one can see from Table 4.3, the particulars in the class do not have a property in common; they are an *imperfect* community and as such we do not want to count them as a property class. (1) and (2) are not sufficient.

The claim that I is not a property class which we want to permit into the system depends upon an assumption which has not yet been articulated and which underwrites the claim that the particulars a, b and c do not share a property. This claim is true only on the presupposition that properties are *sparse*, or relatively sparse, such that combinations of properties are not always properties and, more specifically in this case, that disjunctive properties are not genuine properties. If one were to allow the property of *being F or G or H* to count as a genuine property, then the particulars a, b and c in class I would share a property after all. To adapt the example to a less formal one, let F be the property of *being blue*, G be the property of *being round* and H be the property of *being fluffy*, so that Class I contains one thing which is blue and round, one thing which is round and fluffy and one thing which is blue and fluffy. If we are allowed to presuppose that F, G and H are already individuated, then the simplest property shared between the particulars is *being blue or round or fluffy*; but if we are not even entitled to make that assumption, then the least complicated property shared is that of *being blue and round or round and fluffy or blue and fluffy*, which is a disjunction of conjunctive properties, the sharing of which we are not likely to want to count as the sharing of a genuine property.

Conditions (1) and (2) on resemblance classes have let us down, since they cannot distinguish genuine cases of property similarity from spurious cases exemplified by imperfect communities. Unlike Carnap's strategy in the case of the Companionship Problem, we cannot simply try to resolve the difficulty by assuming that such unfavourable circumstances do not obtain; and besides, the acceptability of such an assumption was highly questionable as a resolution to the Companionship Problem anyway. To make such an assumption, we would require an account of why the resemblance between elements in classes such as I, the imperfect communities, is different to the resemblance between elements in classes such as the class of blue particulars or the class of square ones, and this cannot be provided without presupposing that we already know which the genuine property classes are. This would be begging the question, since on the resemblance class view – which was relying upon (1) and (2) to do the job – we manifestly do not know which the genuine properties, or the 'perfect communities' are.[9]

Goodman's problems show that the conjunction of (1) and (2) is neither necessary nor sufficient to characterize what we intuitively think that a property class should be. If they remain unresolved, they are decisive objections against resemblance nominalism.

4.6 A solution to Goodman's problems?

There have been several attempts to respond to Goodman (Goodman 1977; Lewis 1969: 16; 1983a: 193; Hausman 1979: 199–206; Eberle 1975: 69). Due to the limitations of space, I will not assess these in detail but will focus upon the responses suggested by Rodriguez-Pereyra (2002: 156–85) First, he deals with the Problem of Imperfect Community, about which he notes the following:

> ... what makes α_0 a class all of whose members share some property is that those members resemble each other, and the pairs of the members resemble each other, and the pairs of the pairs of the members resemble each other, and so on. Thus what makes F-particulars have the property F is not merely that they resemble each other, but also that their pairs resemble each other, and the pairs of their pairs resemble each other, and so on. (2002: 171)

What Rodriguez-Pereyra has noticed here is that were there an iterative resemblance relation which held between each member of a pair of the particulars in a set *and* between all of the *hereditary pairs* (that is, between each of those pairs of members and then between the pairs of pairs and so on), the set could not be an imperfect community; in every imperfect community, such pairings would fail at some stage. Accordingly, he introduces the relation R*, a resemblance relation which does hold in this way and which shares the formal features of a resemblance relation in being symmetric, non-transitive and reflexive.[10]

The introduction of R* provides a relation which blocks imperfect communities and removes the danger that the class of all *fluffy and blue or blue and round or round and fluffy* things will count as a property class according to the resemblance nominalists' constraints (1) and (2). But it does so at the cost of introducing a relation which is much more abstract and complicated than the intuitive notion of resemblance with which the account began, and this new relation must be taken as primitive, since it is intended to explain the particulars of a set sharing a property, rather than the other way around. While treating the original relation of resemblance between pairs of particulars as primitive did not seem implausible, the acceptance of this iterative form R* is less so, especially since the iterations must range over classes which include actual and possible particulars. Only if complete coverage of a class and its hereditary pairs is attained is a community perfect, making this solution to the Problem of Imperfect Community epistemologically unsatisfying even if it is ontologically consistent. One cannot look at a local portion of a class – a few actual particulars – and see that R* holds, although one (arguably) can tell whether they resemble each other or not. But such epistemological problems need not carry much weight in the debate here, or even be allowed to enter at all. R* solves the Problem of Imperfect Community if one is prepared to accept it as primitive.[11]

With a solution to the Problem of Imperfect Community in hand, Rodriguez-Pereyra thinks that he can also solve the Companionship Problem. The difficulty in this case is that a class C of particulars which is a subclass of, or coextensive with, another resemblance class fails to count as a resemblance class because it fails to meet condition (2), which states that a resemblance class must be maximal, that is, that no non-members of class C resemble members of C to the same degree as the members of C resemble each other. Thus, an F-class in which all the particulars have property F and also have property G (although not all G particulars are F) will fail to meet condition (2): property classes which are subclasses of other property classes do not yet fit into the resemblance nominalist's account. To fix this problem, Rodriguez-Pereyra introduces *degrees* of resemblance: two particulars resemble to degree n if and only if they share n properties (2002: 179). Furthermore, given that his solution to the Problem of Imperfect Community invokes the relation R* rather than simple resemblance, the relation R* must hold to degree n if and only if the relata of R* (which may be nth order pairs) share n properties.

In our example of the F- and G-classes above, where all the F particulars are also G particulars, but not all of the G particulars are F particulars (such that the F particulars are a subclass of G-class), the F particulars resemble* to degree 2 while the G particulars resemble* to degree 1 (presuming for the sake of simplicity that no other properties are involved in these classes). There is a *lowest degree* to which all the F particulars resemble (in this case degree 2) that is not shared by the G-particulars, which permits us to distin-guish the resemblance relations of the two classes and thereby formulate revised necessary conditions for a resemblance class to be a property class. The particulars of a class must be a perfect community which share a lowest degree of resemblance, and they also must not be a subclass of a perfect community of particulars which resemble to the same lowest degree; that is, the class must be a *maximal* perfect community whose members share a lowest degree of resemblance.

It seems that Rodriguez-Pereyra has managed to evade both the Problem of Imperfect Community and the Companionship Problem,[12] leaving the way open for a viable resemblance nominalist account of the ontology of properties. The resemblance nominalist has a choice about whether to identify properties with maximal resemblance classes which meet the requisite conditions, or simply to use such classes to explain what makes it the case that particulars have the properties that they do (or that we say that particulars have the properties that they do).

However, this success has been achieved with considerable ontological commitment: the relation R*, and the notion of particulars resembling each other to degree n, must both be taken as primitive. In the latter case, the definition of resemblance to a degree is based upon the number of properties shared by a class of particulars – those that share two properties resemble to degree 2, and so on – but, on pain of circularity, we cannot give an account of what it is for particulars to share a property or a number of properties, since properties are the very entities which the resemblance nominalist is trying to characterize.

One might feel at this point that the Companionship Problem has been solved by presupposing that the world contains a solution to it; that is, that the particulars of the world resemble each other by degrees where the individuation of these degrees happens to coincide exactly with the properties that there are. In order to characterize properties in terms of resemblance classes, we have to assume that there are countable layers or strengths of resemblance corresponding to countable increases in the number of properties shared, and yet none of this rich, and seemingly qualitatively individuated, additional ontology is available for further analysis and must be treated as primitive. For some, this solution will come at too great an ontological price, and one might even charge Rodriguez-Pereyra with implicitly reifying properties with his primitive relation of particulars resembling to degree n (Paseau 2012: 377; Morganti 2007).

In light of the ontological commitment involved in using Rodriguez-Pereyra's solution to Goodman's Problems, it is worth considering Lewis's (1983a: 193) proposal to rescue resemblance nominalism. Lewis attempts to solve both problems with one revised resemblance relation and he suggests that a contrastive, variably polyadic relation of resemblance will do the trick; that is, a resemblance class is adequate to be a property class if all the individuals within it resemble each other more than they resemble other particulars. The Problem of Imperfect Community is solved by linking every particular in a class with every other particular, and the Companionship Problem by contrasting these particulars with particulars outside the class which do not resemble them as much. The relation would be something of the following form, let us call it R!:

R! $x_1, x_2, x_3, \ldots$ resemble one another and do not likewise resemble any of $y_1, y_2, y_3 \ldots$

The domains of both the x's and the y's could be of any size, and they could be infinite, and so R! would have to vary in the number of places it had accordingly (hence being *variably polyadic*), it would also have to be treated as primitive. Rodriguez-Pereyra rejects this option (2002: 157), and other suggestions of this type (Hausman 1979: 199–206), since he is concerned that a resemblance relation should be binary; that is, it should only hold between two relata. Given the complexity introduced into Rodriguez-Pereyra's own theory to maintain a binary resemblance relation, Lewis's R! might just be preferable despite its turning resemblance into a relation which holds between however many particulars are in the set (and contrasts them with all the particulars which are non-members). However, as Paseau (2012: 370) points out, as sets of particulars get bigger, Lewis's relation quickly gets out of hand. If we consider R! where both the x_i's and the y_i's are uncountably infinite (for instance, if there are as many x_i's as countable ordinal numbers $\aleph_1$ and as many y_i's as members of the power set of those, $\aleph_2$),[13] then R! becomes doubly uncountably infinite and is a relation which we can only understand in terms of finite concepts of resemblance and our understanding of set theory. It has moved a long way from the intuitive relation of resemblance from which we started.

Preempting such criticisms, Lewis decries his own suggestion, declaring that R! comes at 'a daunting price in complexity and artificiality of our primitive' (1983a: 193). However, this is no loss to him, since his aim is to persuade the reader of the plausibility of his own theory which we left behind when the evaluation of class nominalism was quietly dropped in order to chart the fortunes of the resemblance nominalism. Now we have seen the challenges which resemblance nominalism faces and some suggested solutions, and discovered the extent to which the intuitive notion of resemblance must be complicated while still being kept as a primitive, it is time to return to discuss Lewis's version of class nominalism in greater depth.

4.7 Class nominalism and property naturalness: Lewis's theory

Recall that the class nominalist does not offer to explain why particulars are members of the property classes of which they are members, for he thinks that there is no further explanation to be had. To mitigate the potential difficulties of the coextension problem, on Lewis's modal realist theory a property is a class of actual and possible particulars, and there is nothing more to be said about particulars belonging to the classes they do because *every particular is a member of every possible class of which it can be a member.* But now Lewis is left with an abundance problem as there are too many property classes to be useful as properties: they do not capture resemblance between particulars, there are too many to be causally relevant, the number of them outruns our linguistic ability to formulate predicates for them[14] and they are uncountably infinite. As Lewis states, 'properties carve reality at the joints – and everywhere else as well. If it's distinctions we want, too much structure is no better than none' (1983a: 192). Whereas the resemblance nominalist struggled to avoid imperfect communities meeting his criteria for resemblance classes, the class nominalist has classes for imperfect communities and much more besides, including any collection of particulars which set-forming operations will allow.

To remedy this over-abundance problem, Lewis introduces a primitive distinction between some property classes and the rest. There is 'an elite minority of special properties' which he calls the *perfectly natural properties* upon which the naturalness of other properties depends:

> [A] nominalist could take it as a primitive fact that some classes of things are perfectly natural properties; others are less-than-perfectly natural to varying degrees; and most are not at all natural. Such a Nominalist takes 'natural' as a primitive predicate and offers no analysis of what he means in predicating it of classes. (1983a: 193)

Property naturalness holds as a matter of degree, with classes ultimately deriving their degree of naturalness in virtue of the degree of complexity of their combination from the sparse collection of perfectly natural properties. Thus, imperfect communities such as the class of all *fluffy and blue or blue and round or round and fluffy* things will have a lower degree of naturalness than the class of all *blue* things, or the class of all *fluffy* things; and classes which are so intuitively heterogeneous that they outrun our ability to describe them will be even less natural, until eventually naturalness tails off. If the status of some classes as perfectly natural ones is accepted as a fundamental feature of the ontology, Lewis can have a class nominalist system which includes both sparse properties and abundant properties, and gives an account of how they are related.

Despite the similarity in terminology, property naturalness has nothing to do with Nature or the natural world, or with any other contingent features of the actual world. Property naturalness is not a contingent feature of property classes, nor could it be contingent because such classes can include actual and possible particulars as members. Property classes which are perfectly natural are necessarily so, although they may not have members existing in every possible world (in fact, they are very unlikely to do so), and so the range of which perfectly natural properties there are varies across possible worlds (and must vary, if the worlds are to be qualitatively dissimilar).

Lewis briefly considers the alternatives to treating naturalness as primitive, including having the perfectly natural properties correspond to universals (1983a: 192), or the aforementioned option of developing the account of resemblance classes to avoid Goodman's problems by introducing the variably polyadic relation R!, from which the notion of naturalness could be determined (1983a: 193 n.9). But from the point of view of the class nominalist, the idea of making an additional ontological commitment, even to a sparse ontology of universals along the lines of Armstrong's immanent universals (which is the theory that Lewis has in mind), merely introduces 'idle machinery' where a primitive distinction will do. As Lewis points out, once their task of picking out perfectly natural properties was done, 'the universals could retire if they liked, and leave their jobs to the natural properties' (1983a: 192).

The debate between straightforward class nominalism and the theory which combines class nominalism with sparse universals hinges upon how much one wants to take as primitive, and how much one takes the postulation of additional, almost tailor-made ontology to explain: without universals, the class nominalist takes the existence of some perfectly natural classes of particulars as primitive, but one might wonder whether the existence of universals corresponding to these classes would really serve to explain the objective distinctions between types of things in any more depth. As for resemblance nominalism, that too involves primitive assumptions, most obviously embedded in whichever revised version of the resemblance relation is postulated to guarantee that resemblance captures all and only property classes, thereby fulfilling a comparable role in the resemblance class theory to Lewis's presupposition about the existence of perfectly natural classes. One might even wonder how much difference there is between these theories. As Lewis (1983a: 194 n.9) notes:

> [I]t is not at all clear to me that Moderate Class Nominalism [Class Nominalism with primitive Naturalness] and Resemblance Class Nominalism in its present form are two different theories, as opposed to a single theory presented in different styles.

4.8 Whither nominalism? Classes, universals and tropes

The ontological commitment involved in both resemblance nominalism and class nominalism is considerable if they are to be viable accounts of properties. Both require a commitment to modal realism and with it counterpart theory (unless one takes a very hard-headed approach to the coextension problem), and either a rather artificial and complex reformulation of the resemblance relation must be accepted as primitive, or else a commitment to some properties being more natural than others. The critics of nominalism seize upon these primitive features, citing Goodman's problems and the difficulty of their resolution as additional reasons to avoid resemblance nominalism, and urge the choice of an ontology which does not lead so quickly into trouble.

Moreover, both resemblance nominalism and class nominalism violate the intuition that whether a particular has a certain property is an intrinsic matter which does not depend upon other particulars, since the nominalists think that a particular's having a property is a question of its resembling other particulars, or of belonging to a property class. However, as we saw in 3.2, two of the three accounts of trope resemblance are also nominalist in spirit and suffer from the same defect, so only the universals theorist is in a better position at this point. But the universals theorist has his own trouble with intuitions clashing with his theory, since the universals theories either ask us to accept that an entity can be wholly present at distinct spatio-temporal positions, or that universals are unlocated or abstract, and yet instantiated by spatio-temporally located particulars. It is not clear that one can answer the question of which theory has the greatest intuitive weight.

There are two other considerations to which the nominalist might draw attention in his theory's favour, which concern why the Coextension Problem and Goodman's Problems arise. Although these present serious difficulties and their resolution significantly complicates nominalism, what we have so far neglected to notice is that they arise because the nominalist wants to answer questions which are barely raised within other theories and certainly not given satisfactory answers.

First, the coextension problem arises in the course of characterizing properties in terms of classes of particulars, a project which both requires no specialized ontology to occupy the role of properties or qualities and (if it is successful) will also provide constitutive identity and individuation criteria which determine what makes properties numerically the same or different. As it stands, the universals theory fails on both of these counts: universals almost appear to have been invented to provide the ontology of properties; and we have, as yet, no obvious account of identity criteria for universals. One might complain about trope theory on similar grounds: tropes too are inherently qualitative entities, postulated to answer questions about qualities and as such their postulation may be accused of being ad hoc. Moreover, although an attempt was made to give an account of the individuation of tropes in terms of spatio-temporal location (3.3), this does not individuate tropes qualitatively and so an additional account of trope resemblance is required, an account which is also, in some trope theories, left as primitive. In 5.2, I will discuss whether constitutive identity criteria can be given for properties in general and whether such criteria are required, or at least desirable, in a metaphysical theory. At that point it may transpire that these complaints about the categories of universals and trope lacking specific identity criteria are misplaced, or else that they can be mitigated by the imposition of a more general identity criterion for properties. In the meantime, however, the nominalists can claim to have the upper hand, since property identity is determined set-theoretically in terms of extensions, by which particulars are members of a class.

Secondly, the resemblance nominalist might argue that Goodman's problems are also brought about by the nominalists' enthusiasm for answering questions which supporters of other ontological options have not properly addressed: in this case, the question of why particulars fall into the property classes that they do. Perhaps the fact that each of the range of viable answers to Goodman's problems is complicated and requires a counterintuitive form of resemblance to be taken as primitive should not count against the theory when its competitors can do no better. Both universals theorists and trope theorists explain particulars having the same property in terms of entities which are essentially qualitative and might not even have been postulated had we not been trying to answer the question of what makes particulars qualitatively similar to each other. If nominalists can answer the same question without resorting to essentially qualitative ontology, except for resemblance relations which they can claim to have developed from our intuitive understanding of similarity and difference, then resemblance nominalism remains a strong option among the ontological accounts of properties on offer.

FURTHER READING

Resemblance nominalism: Price 1953; Rodriguez-Pereyra 2002.
Class nominalism: Lewis 1983a.
Objections: Goodman 1972; Paseau 2012; Sider 1996.
See also reading in Chapter 9 for objections to Lewis's account of natural properties.

Suggested Questions

1   Describe the different resemblance structures which a resemblance class can have. Is there anything to recommend the view that the members of a class all resemble a subclass of paradigm particulars?

2   Does the resemblance account of trope similarity avoid the objections to resemblance class nominalism? If so, why? Are resemblance classes of particulars or of tropes preferable?

3   Does the fact that Carnap's two conditions on resemblance classes were formulated to give an account of resemblance between *experiences* make

his account any more plausible than when the conditions are applied to objectively existing resemblance classes?

4   Describe the problems of Companionship and Imperfect Community. Is there a plausible formulation of resemblance nominalism which solves them?

5   Why does Lewis introduce a distinction between perfectly natural properties and the rest, and what does it involve? How plausible is his account in relation to other theories of properties?

6   What role does modal realism play in resemblance and class nominalism? Could plausible versions of these theories be formulated without it?

Notes

1   Nelson Goodman's nominalism resists categorization into nominalism$_1$ and nominalism$_2$, since he attempts to characterize an ontology based solely on individuals, which only excludes universals and abstract entities if these cannot be treated as individuals. His view does rule out the use of classes, however (1977: esp. 24–33).

2   I will take mereological nominalism to be inadequate as a general account of properties. Although it is possible in some cases to identify a property P with the mereological sum, or aggregate, or fusion of entities which have P – the aggregate of blue things is blue, for example – this identification does not work for every property. For instance, the aggregate of all the things with mass of 1 kg is not itself 1 kg, nor is the aggregate of circular things circular; nor do the parts of 1 kg have mass of 1 kg, nor are the parts of a table themselves tables (or, at least, not usually). The mereological account simply does not generalize, nor can it really have been intended as a general account of properties since the objections against it are so obvious. See Armstrong 1978a: 35.

3   For further discussion of the metaphysics of what we normally consider to be impossible, see Yagisawa 1988, Priest 1997, Nolan 2014.

4   The trope theorist who favours resemblance classes of tropes avoids this objection, since her ontology is already distinguished into finer-grained entities, such as individual instances of redness and squareness. By their simple qualitative nature, a trope cannot resemble another trope in more than one 'respect'.

5   In this latter example of determinable and determinate properties, one might just choose to deny the intuition that the properties are distinct.

6   Granger (1983: 22–3, 29–30) and Proust (1989: 192–3) take this view about Carnap's own project with respect to the Companionship Problem, while Rodriguez-Pereyra (2002: 178) claims that Carnap does need an external viewpoint in order to check extensional correctness in his theory. However, it is not clear why Carnap requires, or would be entitled to, such a viewpoint, since for him there are no independent matters of fact to which our similarity sets must match up.

7   See Rodriguez-Pereyra 2002: 153–5 for further discussion of how the problems have been conflated and by whom.

8   If you are concerned that this assumption might not be innocuous, Rodriguez-Pereyra offers a defence of it (2002: 147–8).

9   The term 'perfect community' is Rodriguez-Pereyra's (2002, Chapter 8), who also uses the term 'non-community' for disparate classes in which not all pairs of members resemble each other in some respect or other.

10 The reflexivity of resemblance is slightly counterintuitive (2002: 72–4), since a particular b resembling itself requires the existence of another particular c which it resembles. The same applies to R*. A consequence of this is that no necessarily unique entities resemble themselves.

11 Rodriguez-Pereyra's formulation also requires that no particular has infinitely many sparse properties, in order to generalize his response to cover property classes with infinitely many members. (Otherwise, he has no argument that every infinite imperfect community has some finite imperfect community as a subclass, which is a claim he requires in order that his account does not violate the set-theoretic Axiom of Foundation. To put the point another way, if some infinite imperfect communities are only imperfect in virtue of their being infinite that is, they are such that every finite subclass is perfect then the set theory is not well-founded (2002: 172–4).) I do not think that his arguments for this restriction on how many sparse properties a particular can have are convincing, but it would go out of the scope of this chapter to argue the point here. See Paseau 2012.

12 There are further technical complications in Rodriguez-Pereyra's system (2002, Chapter 11) since some intersections of property classes meet the requirement of being maximal perfect communities whose members share a lowest degree of resemblance, and yet these are not property classes themselves. He calls these classes *mere intersections* and introduces measures to exclude them from the definition. I will not evaluate these amendments and for simplicity I will assume that Rodriguez-Pereyra's solution to this problem works.

13 For the sake of the example, I will presume that the generalized continuum hypothesis holds, such that the power set of a set of cardinality $\aleph_1$ is $\aleph_2$.

14 That is not to say that there will be a property class for *every* predicate (although Lewis suggests that there can be (1983a: 196), presumably as an accidental oversight which he retracts in 2002), since as Russell's paradox shows, some predicates cannot have a determinate extension or class associated with them. For example, the predicates 'is heterological', which means 'does not describe itself', or 'is a set which is not a member of itself' both fail to denote classes.

5

Properties: Grounded or ungrounded? Sparse or abundant?

This chapter compares the ontological theories of properties considered in the previous three chapters and marks out where their differences place them in a broader philosophical picture, and thus which additional philosophical commitments they require or entail. The question of whether constitutive identity and individuation criteria can be provided for properties is considered with the outcome of this matter being found to depend upon whether the ontology of properties is sparse or abundant. Two conceptions of sparseness are discussed: the view that sparse properties are the fundamental ones which determine all the objective similarity and causal interactions in the world, and the conception of sparse properties as those central to the sciences. Schaffer's objection that neither conception is coherent is discussed and rejected. Finally, the prospect of individuating sparse properties in terms of their causal roles or causal powers is considered.

In the previous three chapters, we have investigated the plausibility of several ontological accounts of what properties might be. These fit, broadly-speaking, into three or four different families of theories: properties might be universals; they might be tropes; or they might be sets or classes of particulars, the membership for which might or might not be determined by resemblance relations. The final option, not yet discussed, is that properties may just be properties: they may be ungrounded, primitive entities which are not dependent upon or identical with any other ontological categories.

How is one supposed to choose between these options? The general question of how we should assess ontological theories is a tricky one – and an entire philosophical project of its own – but some guidelines are essential if a principled decision is to be made. Even given such guidelines, it is unlikely that a specific ontological account will be entailed by the reasons which we can give for holding it. Whatever decision is made between theories of properties will most probably rest upon inference to the best explanation: whether an ontological account of properties is better than its rivals depends upon its being more explanatory and its fitting in better with our other philosophical concerns.

Since the assessment of which theory is better in this regard involves the consideration of multiple inter-related philosophical questions, I will not be able to do justice to this project here. Nor, even if I could explore each problem in great depth would I expect to find agreement and definitive arguments in favour of one view rather than another; the emphasis which one philosopher might place upon solving one problem might be outweighed by the importance of an entirely different question when evaluated by someone else. Moreover, the resulting theories will, most probably, all do the job of providing the ontological grounding for similarity and difference which I have been seeking in the previous three chapters. There is much to be said about how we might regard metaphysical theorizing and the status of ontological categories in general, once we have noticed the seeming philosophical equivalence of adopting one ontological account rather than another, but I will not pursue this question here (see Chalmers et al. 2009; Allen 2012). However, with the methodological difficulties in mind, I will lay out the principal considerations which might influence a decision and then leave it to the reader to draw their own conclusions about which ontological account of properties he or she prefers.

5.1 Property theories: The story so far

Which factors count when assessing a theory of properties? In addition to the consistency and coherence of the ontological categories and the relations between them, many of which have been considered in depth already, we might also look towards five broad areas which could be relevant to the decision. First, one might be concerned about the extent to which a theory involves primitive assumptions and how complicated these are: How much must be accepted as brute unanalysable fact in order to make the theory plausible? The fact that a theory makes assumptions which do not admit of further explanation does not by itself count as a disadvantage, since some primitive claims are unavoidable; explanations have to stop somewhere, and the trick is to choose the most plausible place to stop before the assumptions become too convoluted or unlikely to be true. The second, general metaphysical consideration, is whether the categories of entities which are postulated can be identified and individuated: Are there constitutive criteria to say what makes entities of a specific category the same or different? A third issue which has a specific bearing upon the choice which one might make between property theories concerns how one wants to deal with philosophical problems concerning modality. As we have seen, some theories giving the ontological basis of properties involve explicit commitment to a particular view of possibility and necessity, while conversely, a prior commitment to a particular account of what determines possibility might restrict the way in which a property theory is formulated. Fourth, similar, and sometimes related, considerations apply to whether one is prepared to make an ontological commitment to abstract objects, or prefer to be restricted to entities which exist in space and time. The final consideration is whether an ontological theory has greater explanatory power than its rivals, both with respect to whether it can solve other metaphysical problems and how effective it is as an explanation of other philosophical questions more generally.

The theory of universals is successful as a theory of properties precisely because it postulates entities which appear tailor-made to ground qualitative similarity, or types of things; universals do exactly what properties are supposed to do, providing a solution to the one-over-many problem which is a central demand of a property theory. Nor do universals theories explicitly involve a host of primitive assumptions and mechanisms which we are unable to further explain, which puts them at an advantage over the rival trope theories, some of which presuppose ontological mechanisms to ground resemblance and individuation in order to provide the same level of ontological explanation that universals theory offers. But there are two points to note in this regard. First, universals theories are not without primitive assumptions: for instance, we do not seem to be able to give an account of resemblance between universals themselves, nor to explicate the instantiation of universals by particulars without presupposing something like a non-regressive non-relational tie between them. Secondly, one might accuse the universals theorist of simply building assumptions in by postulating specialized repeatable qualitative ontology: we must presuppose that there are such entities as universals – whether immanent or transcendent – in order to avoid the problems which the trope theorist encountered in explicating resemblance relations between tropes. Furthermore, the trope theorist might charge the universals theorist with failing to provide a plausible account of the location of qualities: either the instances of a universal are somehow the instantiation of an abstract entity (on the transcendent view), or a universal is wholly present in each of its instances, and neither seems to be a particularly happy solution to the question of how properties are located in the spatio-temporal world. In contrast, this is a question which the trope theorist answers well, albeit at the cost of significant ontological assumptions about the mechanisms which underlie both the resemblance and the particularity of tropes. Such particularity of qualities might also turn out to be useful in philosophical explanations, again giving trope theory an advantage over its rivals.

However, in contrast to the set-theoretic accounts of properties, neither the Standard Theory of tropes, nor universals theory can provide constitutive identity conditions for properties. The question of what makes tropes on the Standard Theory of the same qualitative type, or what makes universals numerically identical has not been answered in the course of formulating theories about them, while resemblance and class nominalism inherit identity criteria from classes in terms of the concrete particulars or the tropes which are members of such classes. Furthermore, the accounts which stick to an ontology of concrete particulars and classes effectively reduce qualities to non-qualitative entities, and have motivation to postulate the categories they employ independently of explicating qualitative similarity; there is more reason to countenance the ontology they employ than just for characterizing the nature of properties.

On the other hand, both resemblance and class nominalism are committed to some form of modal realism if their identity criteria are to avoid being too coarse-grained, a commitment which may restrict their appeal. Such accounts of properties involve commitment to the existence of possibilia, or one must forego the advantage of property classes having identity criteria in the first place. On the other hand, both the trope theorist and the universals theorist do not require modal realism and can retain an account of modality in terms of actual entities if they choose. Immanent universals are restricted to space-time and permit an actualist account, while transcendent universals do not place restrictions upon what grounds possibility and necessity in metaphysical theories which postulate them.

The decision between the latter transcendent conception of universals or a set-theoretic theory of properties, and immanent universals or trope theory, may be determined in part by commitment to abstract objects (or the avoidance of such commitment). Are non-spatio-temporal entities – either particular or general – coherent? One might think that commitment to such entities cannot entirely be avoided (Quine 1976; Putnam 1979), or conversely that it is possible to keep the metaphysics of the natural world within spatio-temporal bounds. Such questions about modality and abstract objects are too big to be answered here and so I will content myself to flagging up some of the implications which different philosophical opinions about modality and abstracta have for different theories of properties and conversely the implications which specific property theories have for them.

As is now obvious, each theory of the ontological basis of properties has its own advantages and disadvantages, and it may not even be possible to judge whether a specific feature should be counted in an ontological theory's favour except relative to the theory itself, or to other philosophical concerns one might have. One might start to wonder whether it is possible to argue for a univocal position at all. Moreover, since all the theories still in play are viable ways of giving an account of objective similarity and difference, one might start to wonder what the difference between them is. Are they terminological variants of each other? Might they be inter-translatable or be otherwise ontologically equivalent? Or is one theory the correct one despite our present apparent inability to determine which one that is?

One might object that this pessimism comes too soon. Despite the foregoing discussions about the formulation of the ontological theories themselves, one might worry that we do not yet have enough information about the explanatory tasks to which property theory can be put in order to judge which, if any, of the theories is more explanatory. Only once we have examined properties in action can we decide whether one or another theory can do the job better. Accordingly, the final three chapters will be devoted to examining the role of properties in relation to other key philosophical matters including causation, laws, modality, and epistemic questions about how we know which properties there are. Of course, it may turn out that the question of which ontological account of properties is the best one is independent of which characterization of properties is better. But, in order to reach this negative conclusion, we must investigate the uses to which properties can be put in these philosophical contexts and see whether any differences arise.

Before embarking on the project of putting properties to philosophical work, we will look in greater detail at what properties are, how they may be individuated and whether there are any useful distinctions to be made between different varieties of properties. In this discussion, I will set aside questions about the ontological ground of properties (if any) and will treat properties as ungrounded, unless it seems relevant to the discussion to do otherwise.

5.2 Identity and individuation criteria for properties

What are properties? That might strike the reader as a strange question at this point in the book, but it might be sensible to consider it afresh. Examples are easy to come by, and several have been scattered throughout the discussion so far, but it is difficult to get a clear conception of what the members of a category of entities are in very general terms. The preceding chapters have explored different accounts of what properties might be, if they can be understood in terms of other ontological categories; but a second avenue of exploration might be opened up by enquiring what makes properties numerically identical to or different from one another. Can we find a criterion for the identity and individuation of properties?

The provision of such a criterion is sometimes regarded as mandatory, with the failure to find or formulate a plausible criterion being sufficient to rule a category of entities out of the ontology. This stringent restriction can be traced back to Frege: 'If we are to use a symbol a to signify an object, we must have a criterion for deciding in all cases whether b is the same as a' (1884: 62); with the sentiment enthusiastically sustained by Quine (1969). The idea is that we should not countenance the existence of a category of entities without an account of what makes the members of that category numerically the same as each other, or different. In terms of properties, what makes Property P = Property R (say)? Or, what makes properties P and Q distinct? The criterion demanded here is *constitutive*, not epistemic: we are not interested in how or whether we can tell that property P and property Q are distinct (although that would be useful information to have), rather in what it is that makes them distinct, since this might reveal a general fact about their ontological nature.

Other philosophers demure from the strong version of this restriction, however; not least because its rigorous application results in an incredibly limited ontology, with seemingly essential categories of entities being excluded for want of an identity criterion (Lowe 1989). Furthermore, universal adherence to the restriction may even turn out to be self-defeating: for instance, if one maintains that which entities exist is determined by our best theory, then what individuates theories? (Quine 1990a: 99–101; 1990b: 13) Or else, what individuates ontological categories themselves, rather than their members? In addition to, or aside from, the logical problem which might result from its universal application, one might simply judge that this constraint ties metaphysics too tightly to epistemology: Why think that our being unable to discover identity criteria for entities of a certain category makes such entities ineligible to be part of the ontology? We can have a good idea what properties are, for example, without knowing what makes apparently distinct properties the same, or what makes one property numerically distinct from another.

Nevertheless, even though the requirement need not be treated as mandatory, one might still think that identity and individuation criteria are desirable for a category of entities if we can find them, in order to give a clearer conception of the entities under investigation. So, what makes one property the same as another?

Some accounts of the ontological basis of properties will be able to help at this point; in particular, the class or resemblance nominalist theories of concrete particulars or tropes lend set-theoretic identity criteria to properties themselves for those who choose to identify properties with property classes. Sets are identical if and only if they have exactly the same members, which are either tropes or concrete particulars, depending upon which ontology you prefer. Universals and tropes whose qualitative sameness and difference is determined by their intrinsic nature as a matter of primitive fact do not readily provide identity criteria which a property theorist can use. But perhaps the property theorist who favours these ontological options could adapt the set-theoretic approach: property P = property Q iff P and Q are instantiated by all the same particulars. Such a clear criterion was championed by Quine who argued that 'no adequately intelligible standard [for the individuation and identity of properties or attributes] presents itself short of mere coextensiveness of instances' (1981: 183). However, as we noted in 4.4, coextension of actual instances is not sufficient for property identity, since every member of class of particulars might instantiate the same two properties as each other by accident. Such problems lead to the inclusion of possible instances in classes as well and then to ever more fine-grained ways of distinguishing properties. We can place the putative identity criteria in the hierarchy which is illustrated in Table 5.1.

The move from properties being identified and individuated by their actual instances to the fine-grained account, in which a property's possible instances are taken into account as well, requires a commitment to independently determined necessity and possibility. For the set-theoretic accounts of properties, this must be done in terms of a realist ontology of modality. However, given another characterization of necessity, one could maintain an alternative account of properties and yet still claim that properties are identical if and only if they are necessarily coextensive; the fine-grained account would still be coherent. But one might think that this amendment is still insufficient and does not distinguish properties finely enough. Some mathematical or geometric properties appear to be *cointensive* (i.e. necessarily coextensive) and other properties appear to be trivially so, simply because they are instantiated by every possible entity. The latter species of

Table 5.1 The Identity and Individuation of Properties

	P = Q if and only if	Counterexamples/ Problems
Coarse-grained individuation Actually coextensional properties are identical	P is instantiated by the same actual particulars as Q	The coextension problem (4.4)
Fine-grained individuation Necessarily coextensional, or cointensional properties are identical.	P is instantiated by the same actual and possible particulars as Q	Necessarily coextensive properties; e.g. *being triangular, being trilateral; being such that a square exists, being such that 2+2=4*
Ultra-fine-grained individuation Hyperintensionally individuated properties	P is linguistically or syntactically identical to Q	*being blue and triangular* is distinct from *being triangular and blue*

properties are known as *indiscriminately necessary* properties for this reason and include *being self-identical*, and *being such that the number seven exists*. The 'fine-grained' account is simply not fine-grained enough to distinguish between such properties and individuate them, we must leave behind the idea that differences between properties implies separability of their instances; properties may be distinct even though they *cannot* be instantiated by distinct particulars.

This move to an ultra-fine-grained account creates problems for those who were hoping for a set-theoretic account of properties like those examined in Chapter 4, unless the domain of worlds postulated by the modal realist includes 'impossible' worlds as well as all the possible ones. For example, we require a world in which at least some trilateral closed shapes are not triangular, or vice versa. This view is not without its supporters (Yagisawa 1988), but the requirement that impossible particulars exist in the same sense as possible ones makes it difficult to see how we can sustain parity between possible worlds and the entities they contain and the impossible ones. Moreover, even if this parity thesis can be sustained, the intuitive attraction of identifying properties by their having exactly coinciding instances is lost if the particulars which can instantiate them include *round squares* and triangles which lack three sides. Accordingly, most hyperintensional versions of property individuation give an alternative account of how *being triangular* is a different property from *being trilateral,* and are prepared to drop set-theoretic identity conditions and the advantages of simplicity and clarity which these bring. Instead, one might opt for a linguistic or a purely formal account of property identity: properties are the same if they are syntactically identical and distinct if not. This produces incredibly fine-grained distinctions between properties in which, for instance, the property of *being wooden and brown* is distinct from the property of *being brown and wooden*. Alternatively, one might try to individuate properties by recourse to abstract objects (Zalta 1983, 1988) or to objectively-existing concepts to provide an ultra-fine-grained conception of properties which outruns the distinctions which the entities of possible worlds can easily provide. But such views also require additional ontological commitment to abstract objects or concepts and, if the individuation of these entities relies upon properties, then an ontological circle results.[1] At present, there is little agreement about what the 'hyperin-tensional' criterion of property identity amounts to, except that the resulting ontology is finer grained than that provided by necessary coextension; more research is required (see Jespersen and Duži 2015).

Do any of these criteria capture our intuitive conception of properties? The moves from an extensional set-theoretic criterion to an intensional one, and then to an even finer-grained, hyperintensional account, were proposed in order to ensure sufficiently fine-grained distinctions between properties, but now one might complain that we have more properties than we need. Our intuitive judgements have been concealing the fact that an important question has yet to be addressed about what the optimum number of properties is. It would be prudent to investigate further before more suggestions for identity criteria are evaluated.

The question of how many properties there are does not call for an exact answer, nor even a numerical one; even if there are comparatively very few properties, there may be very many of them, even infinitely many. For instance, if different quantitative values of a determinable count as separate properties, then different masses such as *being 1 kg*, *being 4.2 g* and *being 3.4 kg* all count as distinct properties, rather than being one property, *mass*. The key factor in deciding what the optimum number of properties is concerns which role properties occupy in the ontology, what they are to be 'used' for, or what they are thought to do. Only once we have some idea of this, can we say whether we want to endorse such fine-grained distinctions that a purely syntactic difference between predicates changes the property referred to, or whether a coarser grained ontology is preferable. I will consider this matter in the next section.

5.3 Sparse and abundant properties

It seems intuitively obvious that there are too many properties if any set of particulars, or ordered pairs, or n-tuples, counts as a property or a relation – or else, on a trope ontology, if any set of tropes or ordered pairs of tropes is equivalent to a property – but it is less clear what the optimum number of properties should be. This difficulty afflicts class nominalist accounts of property theory (3.2.3; 4.7) and in that case, it was dealt with by making a primitive assumption that some classes are ontologically superior to others. A similar concern arose in the more general discussion of properties above as it became clear that evaluating whether an identity criterion determines an ontology which is finely-enough-grained and yet not-too-finely-grained cannot be answered in isolation without some idea of what the theory of properties is for. For instance, if one thinks that properties determine the meanings of predicates (especially if we presuppose the existence of a property to correspond to each predicate), then there will be more properties than if one thinks that the narrowest possible range of properties exists in order to determine the fundamental nature of everything, or all the causal happenings of the world. The former properties have gained the name 'sparse' properties and the latter 'abundant' ones.

'Abundant' is being used as a comparative term here: although properties acting as individual semantic values for predicates will (most probably) be more abundant than fundamental causal properties, there are still far fewer semantic values than there are sets of particulars which could be identified with properties on an unrestricted class nominalist account; that is, before any primitive ontological features, such as property naturalness, are postulated to distinguish some sets from others to keep the population of properties down. I will reserve the term 'super-abundant' for the over-populated ontological extreme, but some philosophers use 'abundant' to describe the same realm (Lewis 1986: 59; Sider 1996). Alternatively, one can talk about sparse and abundant conceptions of properties as the *minimalist* and *maximalist* conceptions respectively (Swoyer 1996).

These different conceptions of properties might represent genuine ontological alternatives: one could support an abundant ontology of transcendent universals, for instance, on the grounds that it will help to provide an account of the meaning of predicates, and more besides (since there will be universals for which we do not have predicates); while someone else might think that the properties which exist are sparse, spatio-temporally located entities which determine causal interactions, with meanings being accounted for in a different way (if at all), perhaps by the postulation of alternative entities such as concepts. Since the conceptions are not directly in conflict with each other, there would be no inconsistency in not choosing between them: one might opt for a dualistic ontology in which sparse properties occupied one role (that of determining causal interactions, say) and abundant ones determined meaning. A more perspicuous version of this approach – and arguably a more elegant one – would seek to give an account of how the two conceptions of properties are related to each other, and thus belong to only one ontological category after all.

Lewis's solution, as we saw in 4.7, is to distinguish some sets from the super-abundant collection as being sparse, perfectly natural properties, and other natural properties are generated by set-theoretic combination from these. The naturalness of each property (other than the perfectly natural ones) is a matter of degree and, at some point of complexity, naturalness tails off. Thus, *being negatively charged* will (probably) be more natural than *being green*, which will be more natural than *being grue*, which in turn will be more natural than *being less than fourteen metres away from a grue elephant in springtime and eating pizza in the rain*. The sparse properties thereby determine the existence of a sufficiently abundant collection of natural properties to account for 'higher level' or non-fundamental scientific properties, for meaning, the semantic content of intentional states, and other uses besides.

5.3.1 Two conceptions of sparseness

Lewis's ontological account of sparse and abundant properties is unified, but his objective distinction between perfectly natural properties and the rest is a primitive distinction, and therefore not so different to Armstrong's claim that some universals are *genuine* – objectively more fundamental than the others – or similar claims which might be made about a subgroup of kinds of tropes. Thus, differentiating sparse properties from the others is not a move which is limited to class nominalism; perhaps only the resemblance nominalist can give a non-primitive reason why certain properties or property classes are more fundamental or natural than others because he can cite the resemblance relation which holds between all the members of a particular class as grounding their shared qualitative nature. But unless he is content with abundant properties, the resemblance nominalist will have to make a primitive assumption about why some resemblance relations are more fundamental than others; why it is that the resemblance between green particulars (say) is more ontologically basic than resemblance between grue ones. So far, none of the sparse property theories have given individuation criteria for sparse properties and have instead relied on a primitive ontological assumption. In this, the different ontological accounts of properties are at least all equal with each other and we may yet discover a generic non-primitive basis upon which sparse properties are distinguished from the rest.

However, in the absence of an identity criterion and with no more than an ontological presupposition that sparse properties exist, we might wonder which properties are the sparse ones: are they the fundamental building blocks of the world; or are they the properties required and invoked by our scientific understanding, drawn from many 'levels' of reality, such as biology, chemistry, psychology, geology and so on? Following Schaffer (2004), we could call the former the *fundamental* conception of sparse properties and the latter the *scientific* one. So far the discussion has largely presupposed that the most desirable conception of sparse properties is the fundamental one, and both Armstrong's and Lewis's accounts take this line. But one might question whether this presupposition is justified in view of the ontological roles which we expect sparse properties to play.

It would be desirable for the members of the category of sparse properties to fulfil three principal roles: minimality, similarity and causality. Or, to put the point another way, sparse properties are the properties which carve at nature's joints; they determine objective similarity and causal interactions with as minimal an ontology as possible. However, Schaffer (2004) suggests that these roles conflict, with the scientific conception of properties more suited to marking similarity and causality, and the fundamental conception tailor-made to conform to the minimality constraint. He gives examples of intuitively non-fundamental similarity and causality to make his claim: surely two conscious subjects who have the desire to get inside from a rainstorm (desire R) have something – a property of desiring R – in common and if their beliefs and desires cause them to act and to run towards the nearest shelter, then the properties in virtue of which they are acting are the shared desire R and their beliefs that the shelter will get each of them out of the rain. Such causal interactions between what we think and what we do form an important starting point for many debates in the philosophy of mind; as Jerry Fodor remarks, 'If it isn't literally true that my wanting is causally responsible for my reaching ... then practically everything I believe about anything is false and it's the end of the world' (1990: 156). Neither the shared desire R, nor the associated beliefs, are fundamental sparse properties, but they do mark objective similarity and difference between people and they play a causal role. Furthermore, even if the psychological properties in question are determined by fundamental sparse properties, they are likely to be multiply realized by them; that is, each psychological property (R, say) is determined on different occasions, or in different subjects or species, by different physical properties and so the similarity at the psychological level is not present at the level of fundamental sparse properties. Although fundamental properties are minimal, as is required of sparse properties, they do not satisfy the criteria of marking similarity, nor of determining causality. Similar remarks would apply to other properties which do not obviously fit into the minimal sparse set of fundamental properties, such as macro-physical properties, chemical properties, biological properties and the like, so that even if one had qualms about the example of mental causation presented above, there would be others to replace it. It seems plausible to suggest that there is more similarity and causality than just that in micro-physics.

A supporter of the fundamental conception of sparseness could give two, potentially compatible responses to this objection. First, he could bite the bullet about the real or true objective nature of similarity and causality: the ultimate nature of the world is determined by properties at the fundamental level, and higher level properties – especially those which are multiply realized – do not cut nature at its joints. Although the properties of beliefs, desires, jade, rainforests, wombats and terminal moraine are useful in our explanations, they mark similarities and causal interactions which we find interesting and useful to pick out, not those which are of the greatest ontological importance insofar as causality and similarity are concerned. Thus, many properties of science do not fulfil the criteria of being sparse, even though our explanatory experience makes it appear that they do. On this view, the properties used in science have an explanatory role to play: they cannot be eliminated in favour of the sparse fundamental properties since to do so would replace heterogeneity with homogeneity in our explanations. For instance, we can give a better explanation of why a 5 cm square peg will not fit into a 5 cm round hole using properties instantiated by ordinary middle-sized objects, rather than those of sub-atomic physics, because the former properties carve nature at the level which is relevant to our explanatory aims (Putnam 1975a: 295–8). But that, one might argue, is a pragmatic feature and does not make such properties part of the sparse ontological elite which form the basic building blocks of the world.

Second, the supporter of the fundamental conception could partially rehabilitate the ontological status of scientific properties by maintaining that they are related to fundamental sparse properties. On Lewis's system for instance, non-fundamental scientific properties are natural to quite a high degree, while Armstrong considers the universals used in science to be related to simple or genuine universals because the former are conjunctions, or structural assemblages of the latter (Armstrong 1997: 31–8). Although tenable, this response requires a strong fundamentalist assumption about the structural relations between properties in the natural world, in particular that all properties are determined by the sparse, fundamental ones and that higher level properties determine similarity and inherit their causal powers in virtue of them.[2] But this belief is a core motivation behind the postulation of fundamental sparse properties in the first place; the belief in sparse properties and the ultimately unified nature of the world go hand in hand.

There are two immediate difficulties with the second suggestion. First, it is not clear whether it is plausible to assume that fundamental properties determine the causal powers of the rest and fix higher-level relations of similarity and difference. Ontological fundamentality may not be sufficient to determine all the similarity and causality that there is. For instance, a higher level property S may be determined by different lower level properties when S is instantiated in different particulars, or in the same particular at different times. Although the existence of each of S's instances is determined by some fundamental property or other, each instance might be determined by a different fundamental property; S is multiply, or variably realized by fundamental properties. Thus, instead of S being neatly correlated with a fundamental property, or a collection of fundamental properties, S would be correlated with a disjunctive set of such properties, only one of which need be instantiated for S to occur. For instance, if psychological properties (say) are multiply realized, they are correlated with disjunctions of physical properties. Such disjunctions may be heterogeneous on a physical level; that is, they may not mark out any physical similarity at all when the properties in the disjunction are considered together. S is a higher level property, but not a physical one, and so we would not want to say that S determines similarity between the particulars which instantiate it in virtue of the qualitative natures of the heterogeneous collection of physical properties which it ontologically depends upon.

Furthermore, disjunctions of fundamental properties may be open-ended, perhaps even infinite. For example, the economic property of *being a monetary exchange* can be instantiated by paying with coins, or notes, or cheque, or card, or with bitcoins, or by another electronic transaction, or by bartering with beads, or swapping a horse for equal value of goods, or paying with labour (by washing up to pay for a meal in a restaurant, say) and so on. Even since this example was first presented by Fodor (1974: 103), the means by which one might pay for goods or services has expanded an unpredictable way. Such special science properties are not homogeneous at a lower, fundamental level and so it is implausible to think that special science properties determine higher-level similarity if they can only inherit this status from the more fundamental property which determined them on each specific occasion. If we take such examples of multiple realizability seriously, then we must either deny higher-level properties their claim to determining genuine similarity and with this (most probably) call into question their explanatory utility too, or we must abandon the broadly reductive, unifying claim which was invoked to uphold the fundamental conception of sparseness.

Moreover, if the properties used in science are determined by sparse fundamental properties, then this presents philosophical difficulties concerning their causal efficacy. If the fundamental sparse properties determine everything that happens in the world, then they appear to exclude other properties from being causally efficacious. Although it appears that my thoughts cause my actions, the causal efficacy of the psychological properties is preempted by more fundamental, most probably physical, properties which determine the existence of those psychological properties.[3] If we maintain a strong broadly reductionist account of how fundamental properties determine the scientific ones, it becomes implausible to think that scientific properties determine similarity and furthermore they seem to lose their causal efficacy too. For those who are happy with sparse fundamental properties taking all of the qualitative and causal glory, this reductionist account of what occurs in the world will be acceptable; non-fundamental properties can maintain some pragmatic explanatory utility but objectively they will not be doing serious ontological work. However, one might consider the loss of genuine causal efficacy and similarity in non-fundamental properties to be too counter-intuitive to be acceptable and there is much debate about whether their causal efficacy can be reinstated while keeping the conception of sparse fundamental properties in play. On the other hand, those who are prepared to jettison the fundamental conception of sparse properties, and the claim that these determine the rest, are faced with the question of whether the sparse scientific properties which there are in the world fit together into a unified whole. It is impossible to do justice to this problem here, but there are difficulties for scientific properties whether they are independent of the fundamental base or ontologically determined by it.

Furthermore, even if the supporter of the fundamental conception of sparse properties can provide a mechanism by which more abundant, 'higher order' properties are determined by fundamental properties, one might be concerned that the minimality constraint can only be met if there is a fundamental level of reality. But it seems possible that the world could be infinitely complex – there may be no fundamental level of reality in the way we are accustomed to thinking about it, at the level of the properties of physical particles, for instance – and nature might just continue to subdivide ad infinitum. In this case, there will be no fundamental level and no way of satisfying the minimality constraint and so the primary advantage of the fundamental conception of sparseness will be removed. Only if we rule out such infinitely complex worlds can fundamental sparse properties be rehabili-tated, and one might object that such a ruling about the nature of reality amounts to an ad hoc assumption.

Some proponents of sparse properties are prepared to accept the restriction that if nature is infinitely complex, their fundamentalist account of sparseness does not apply; or, alternatively, if there are many possible ways that nature could be – that is, if modal realism is true – then there are sparse fundamental properties only in those possible worlds which have a fundamental level.

For instance, Schaffer reports that Lewis is prepared to treat worlds which are infinitely complex as unlikely enough to admit them as exceptions to his account (2004: 97). In infinitely complex worlds, some properties are more fundamental than others, but there is no basic, finite sparse set upon which all the others depend.

It seems, therefore, that the class of properties which counts as sparse depends upon other views which you have about the nature of the world: specifically, whether one holds a broadly reductionist account of relations between properties such that abundant 'higher level' properties are completely determined by more fundamental ones, with all claims to higher level autonomy being an explanatory, rather than an ontological matter; and, secondly, whether there is a fundamental level in the first place. If these claims are acceptable, then the fundamental conception of sparse properties is the one to choose. If not, then one must either challenge the conclusion that the fundamental conception threatens the autonomy of non-fundamental properties in determining causality and similarity, or else one must embrace the scientific conception of sparseness, in which sparse properties are drawn from all 'levels' of discourse, and the metaphor of levels should ultimately be thrown out. On the scientific conception, sparse properties are just those which primarily account for similarity and causality in our best scientific explanations and they do so because they articulate the qualitative structure of nature: properties of the 'special' (probably non-physical) sciences would have no lesser ontological status than those of fundamental physics.

5.4 Another perspective on individuation

In light of the discussion about sparse and abundant properties, a new criterion suggests itself for individuating properties; namely, that properties might be individuated by their causal role, by what they can do. Such a criterion would individuate sparse properties and whichever other properties are causally efficacious, although it would not do for those properties which do not enter into causal interactions at all. Thus, we might suggest that a suitable criterion for property individuation is the following:

(R) Property P = Property Q *iff* P and Q have the same causal role, and are distinct otherwise.

In this case, the causal role of a property will have to be understood in terms of all the causes and effects which that property could have in different situations, rather than what it actually does. As such, we may want to describe this as the property's nomological or nomic role – its role in generalizations or laws – rather than just as its causal role, although I will stick to using 'causal role' for the meantime, since the notion of a natural law has not yet been considered.

This suggestion cannot simply involve individuating properties by what they actually do because the context in which a property is instantiated might serve to impede its causal power, or the requisite conditions for a causal interaction may never arise. An individual particle with negative charge repels other negatively charged particles and is repelled by them, but this will not occur if it is never sufficiently near another negatively charged particle. But we do not want to conclude on the basis of this contingent circumstance that the isolated particle does not instantiate the property of having negative charge. The causal role which a property has must amount to more than what it actually does and encompass what it can do and would do; instantiating a causally efficacious property such as *having negative charge* involves instantiating an entity which has some inherent modal force.

What that modal force amounts to will vary depending upon what one thinks the relationship is between a property and its causal role. If properties have their causal roles contingently, the criterion under discussion will be of limited use: it will serve to distinguish actual, instantiated properties from each other but it will not determine the nature of those properties in any deeper metaphysical sense; what that property can cause will be a poor measure of what makes it the property that it is, except within a specific possible world or situation. On the other hand, if the causal role of a property is necessary, or essential to it, then the causal role criterion will identify and individuate that property in all possible situations. Properties on this latter view will be essentially causal entities, and this might give us a useful and interesting indication of what a sparse, or fairly sparse property is. An intermediate view between these extremes would cast properties as having their causal roles as a matter of nomological necessity; that is, a property retains its causal role across all those possible worlds in which the laws of nature are the same, and may have a different causal role where the laws of nature are different. As in the case of contingent causal role, this criterion will have limited metaphysical import about the nature of properties themselves, although it would serve to identify and individuate properties across nomologically similar worlds. Causal role will not provide necessary and sufficient conditions for a property being the property that it is, if the causal role of a property is not necessarily or essentially related to it. So the existence of causally efficacious properties which have their causal roles contingently will require alternative individuation criteria, if they have individuation criteria at all. The question of what the modal strength of the relationship is between a property and its causal role will be considered in Chapter 7.

Another caveat is that, if causal individuation works at all, it will obviously only work for properties which are causally efficacious; any properties which are not will require alternative criteria, or we will have to make do with the criteria we already have. For example, the properties of abstract entities, such as *being a thrice-instantiated transcendent universal,* and probably also mathematical properties such as *being prime,* or *being divisible by thirteen,* do not enter into causal relations of any kind and so have no causal role. In addition, some philosophers argue that spatio-temporal properties are causally inert and thus these cannot be individuated causally either (Ellis 2000: 331 n.3). If properties are best individuated by their causal roles, the existence of these families of causally inert properties will render such a criterion less than general, and so causal power will not be constitutive of all entities which belong to the category of properties. The causal criterion may, however, still be sufficient to identify and individuate sparse properties, depending upon whether mathematical properties are sparse, and whether spatio-temporal properties which determine position and location can be causally individuated and whether they too are deemed to be sparse or not.

In addition to properties which entirely lack causal connections with other properties, there might also be epiphenomenal properties; that is, they are caused by other properties but they are not themselves causally efficacious. For instance, on some accounts, psychological properties, or some properties of consciousness, such as qualia or phenomenal properties are thought to lack causal efficacy. Unless such properties have unique causes by which they might be individuated, these too would fail to be covered by the causal criterion of property individuation.

One might also wonder how closely criterion (R) phrased above in terms of causal role is related to one couched in terms of the causal power of a property, such that:

(P) Property P = Property Q *iff* P and Q have the same causal power, and are distinct otherwise.

A causal power criterion, as opposed to one based on causal role, would individuate and identify properties in terms of their effects alone, what it is that they can cause, and thus will exclude properties (if there are any) which can only be individuated by their causes. This would preclude epiphenomenal properties which are distinguished by (R) from being covered by (P), as well as any other properties which differ only in what causes them and not what they can cause. Other than these properties, the taxonomy of properties which will be provided by these two criteria will be the same, and they are rarely distinguished in the literature. As we will see in Chapter 7, for some philosophers, properties are essentially causal powers, or at least some properties are. If one takes this view about the essential nature of properties one might think that the key difference between causal power and causal role is that the causal power of a property is intrinsic to it, whereas a property's being determined by its causal role – which causes and effects it has or could have – involves extrinsic features, namely its causal relations to other properties. However, if one takes the role of constitutive identity criteria seriously in metaphysics, this difference is merely one of perspective: if properties can be individuated or identified by their causal role, then this tells us something about the intrinsic nature of properties even though the causes and effects which a particular property has are extrinsic to it.

FURTHER READING

On identity and individuation: Lowe 1989; Achinstein 1974; Shoemaker 1980.
On sparseness: Swoyer 1996; Schaffer 2004.
On ontology: Hirsch 2002; Bennett 2009; Benovsky 2014.

Suggested Questions
1   How can we, and how should we, evaluate a metaphysical theory?
2   How different are the ontological accounts of properties which have been considered in the preceding chapters?
3   Explain the different roles which sparse and abundant properties could play in philosophy. Is one conception preferable to another? If so, why? If not, why not?
4   Is there a coherent, unequivocal conception of sparse properties?
5   What are the most plausible constitutive identity and individuation criteria for properties?
6   Which account of the ontology of properties considered in Chapters 2–4 do you find the most plausible, and why?

Notes

1   This suggestion requires that abstract objects (or concepts) be individuated independently of properties if it is to work, or it may turn out to be circular. I will not evaluate these suggestions here.

2   There are different accounts of what this determination amounts to, including supervenience, realization, or composition of higher-level properties by fundamental ones, but I will not explore these variations here.

3   See Kim 1998 for an overview of this problem, to which there have been many attempted solutions. See also Heil and Mele 1993, Bennett 2003.

6

Intrinsic and extrinsic properties

Is there a coherent distinction between intrinsic and extrinsic properties? Several attempts to draw the distinction are investigated, including metaphysical criteria based on duplication and perfect naturalness, naturalness, and grounding, and logical criteria based on relations to other particulars. The criterion based on perfectly natural properties is found to have limited appeal due to its ontological commitment, but I argue that the grounding account also requires such commitment to avoid circularity or regress. I investigate the attempt to define extrinsicality and argue that this account is more plausible if it permits properties involving abstract objects to be intrinsic. I examine whether the distinction conflicts with the criterion of property identity based upon necessary coextension. Finally I consider whether we are dealing with one distinction or several, arguing that there are two related conceptions of intrinsicality, and one based on property ascription which is narrower than the rest.

When I consider one of the books in front of me – a copy of Lewis's *On the Plurality of Worlds* – it has plenty of properties. It has a mass of 640 g, weight of 6.27 N, the cover is dark red, it was written by David Lewis, it belongs to the Oxford University Library, it already cost me forty pence in library fines (now paid), it is within four miles of a chicken hut, it is cuboid, it is not spherical, it is highly regarded among philosophers (whether or not they agree with its content), it is in English, it is made of paper, board, glue and ink, it is underneath Bas van Fraassen's *The Scientific Image,* and its slightly yellowed pages smell faintly of cigarette smoke. Some of these properties are ones which the book has in virtue of its own nature, independently of whatever else may exist in the world, while others require the existence of other entities in order to be instantiated. For example, consider the distinction between the book's mass and its weight: while the book's mass is independent of its environment and the other entities which exist, its weight changes according to the strength of the gravitational field in which it is located; *weight*, unlike *mass*, is dependent upon the environment. Similarly, being a certain shape – *being cuboid*, for example – is a property which the book can have regardless of the entities which coexist with it, while *being underneath The Scientific Image*, or *being within four miles of a chicken hut* are not.

Intuitively, we make a distinction between intrinsic and extrinsic properties; those which a particular has in virtue of the way that it is and those which require the existence of other entities in order to be instantiated. This distinction has a range of philosophical uses. If I wanted to produce a duplicate of the particular book, the duplication of some properties – such as its colour, shape, the arrangement of ink on the pages – would matter, while others, such as its location relative to the works of Bas van Fraassen or to a chicken shed, would not. Moreover, if I were to hurl the book across the room in a fit of philosophical frustration, it would be in virtue of certain properties and not others that the flying hardback did damage to my surroundings, perhaps (if I were unlucky) breaking a window with its momentum and comparatively robust form. As far as its physical causal efficacy goes, its being written by David Lewis makes no difference to whether shards of glass from my window are now strewn outside on the ground, while if we were interested in its efficacy in generating ideas among other philosophers, its authorship and the semantic values of the words and phrases within it would be of paramount importance.

Until fairly recently, philosophers have been content to rely upon an intuitive distinction between intrinsic and extrinsic properties, but a surge of interest in clarifying what exactly the distinction amounts to has led to a corresponding surge in proposals about how it might be drawn and also raised questions about whether one distinction is being drawn or several. In this chapter, I will survey some (but not all) of the notable alternatives and examine whether there are different notions of intrinsicality in play.

6.1 Navigating the problem

Before evaluating specific proposals, it will be useful to make some general observations about the way in which one might approach the project of distinguishing between intrinsic and extrinsic properties. First, we can roughly divide the different strategies to the distinction into two: those which attempt to draw it on broadly logical grounds, using notions from logic, set theory, possible world semantics, mereology and the like; and those which also rely upon a primitive metaphysical notion, or some kind of ontological commitment, without which the proposal will not work. The second, more complex issue concerns the different families of properties with which we will be concerned. These differences are important since some of the proposed distinctions vary in the scope of which properties they apply to, and others principally disagree about which side of the distinction a certain family of property should fall. But before these are discussed, a caveat or three is needed.

First, I should note that almost any of the intuitive examples given of intrinsic, or extrinsic properties may turn out to be inadequate. For example, one can argue that neither mass nor shape are intrinsic, contrary to how I have presented them. In the former case, this might be because the mass of a particle is determined by its relation to the Higgs field (Bauer 2011), or because mass is essentially related to other properties in a law-like way (Mumford 2004). In the case of shape, there are two problems: first, that the shape of holes may present an exception to the claim that shape is intrinsic, since the shape of a hole is determined by the shape of the material surrounding it; second, a more generalized claim could be made that shape is determined by the properties of space-time, rather than the properties of the thing which has the shape. These are interesting issues, but resolving them need not concern us here, except to note that nothing in the arguments which follow will depend upon whether a specific property used as an example is intrinsic or not.

Second, some properties could be considered *partially extrinsic,* and thus also partially intrinsic. For instance, *being a widow* is extrinsic while the woman concerned is in ignorance of her widowhood, but once she learns about it, it will be an aspect of her intrinsic nature too. For the purposes of this discussion, I will count partially extrinsic properties as extrinsic, unless the difference matters.

Third, while *mass* is a good candidate for being an intrinsic property in every individual which instantiates it, some properties might count as *locally* intrinsic. For example, *being such that there is a dog* is intrinsic when instantiated by a dog, but extrinsic when instantiated by anything which is not a dog (excluding anything, should there be anything, whose existence requires the existence of a dog). Thus, we can draw a distinction between properties which are *globally intrinsic* – that is, intrinsic in every particular which instantiates them – and those which are *locally* intrinsic or extrinsic. We might otherwise say that properties can be instantiated in an intrinsic or an extrinsic fashion, or that properties can be instantiated intrinsically or extrinsically in different individuals, such that a globally extrinsic property *such as being such that there is a dog* can be instantiated intrinsically in certain cases, namely by dogs and by entities which require their existence (Figidor 2008). Unless stated, I will restrict intrinsic properties to those which are globally intrinsic.

There are several families of properties which frequently appear in discussions of the different versions of the distinction. For simplicity and ease of reference, it will be convenient to summarize them on Table 6.1.

Table 6.1 Some varieties of properties

Property name(s)	Description	Examples
Qualitative Properties, Pure Properties	General qualities	being red, having negative charge, being an elephant, being near a zoo.
Existential Properties	Property which requires the existence of something or other (usually of a certain type)	being such that a sphere exists, being such that a bear exists.
Haecceistic Properties, Identity Properties, Impure Properties	Property involving a particular entity	being Obama, being 200 km from Dhaka, not being written by David Lewis.
Identity Properties	Subset of haecceistic properties involving *being* a particular thing	being Obama, being Marie Curie.
Indiscriminately Necessary Properties	Property instantiated by every particular	being such that 2+2=4, being self-identical, being such that a triangle exists.

6.2 Why make the distinction? Some philosophical uses for intrinsicality

The distinction between intrinsic and extrinsic properties features in ethics, aesthetics, epistemology and the philosophy of mind, in addition to metaphysics. The following is a small selection of the philosophical purposes to which it is put.

1   *Change.* As an oak tree grows from an acorn its intrinsic properties
    change; it changes mass, size and shape, but also its internal structure
    becomes more complex, with different parts developing distinct
    chemical and biological properties and functions. The tree also changes
    when someone takes its photograph, or when a bungalow is built a
    kilometre from it, or when a tree preservation order is placed upon it,
    but in these latter cases, the change does not affect the tree itself in
    the same way as in the former ones. We can distinguish between *real
    change* – change which occurs in the intrinsic properties of a thing
    (although it may also include changes to extrinsic properties) – and
    merely Cambridge change which solely involves changes in the extrinsic
    properties of a particular, rather than changes in internal constitution.[1]

2   *Supervenience.* We might try to capture dependency relations
    between different species of properties by saying that the properties
    of one species supervene upon the properties of another. For
    instance, some physicalists claim that the mental properties of an
    individual are determined by the physical properties of that individual,
    such that there can be no change in mental properties without a
    difference in physical properties, or that physical duplicates would
    have identical mental properties. But some properties seem to be
    more relevant than others; the intrinsic physical properties of an
    individual's brain and body appear to be more important to which
    mental properties she instantiates than her being five miles from a
    wooden boat, or being ignored by a distant bear (Kim 1982, 1993b).
    It has also been suggested that moral properties supervene upon
    natural ones, aesthetic properties supervene upon natural properties,
    or biological properties supervene upon chemical ones, which in turn
    supervene upon physical properties.

3   *Internalism versus Externalism about mental content.* Is the semantic
    content of psychological states such as beliefs and desires entirely
    determined by the internal, or intrinsic, properties of the individual
    who has them? Internalists, or individualists, about mental content
    maintain that it is, while externalists have presented arguments in
    which intrinsically identical individuals have different mental content
    in virtue of differences in their environments (Putnam 1975b; Burge
    1979). This matter also influences whether it is plausible to regard
    mental properties as being supervenient upon the intrinsic physical
    properties of an individual, as in (2) above; although the truth of
    externalism does not rule out psycho-physical supervenience claims,
    since the base of relevant intrinsic properties can be broadened to
    include some properties of the individual's environment.

4   *Intrinsic and Extrinsic Value.* We can draw a distinction between a
    thing or an action's being valuable or good in itself because of its
    intrinsic properties, or being valuable because of its relations to other
    things, perhaps because people regard that thing as being valuable,
    for instance.

5   *How do entities persist? Perdurance versus Endurance.* The 'real
    change' described in (1) involves changes in the intrinsic properties of
    a thing or substance, but how are we to understand the persistence
    of that thing over time if its intrinsic properties alter? This is Lewis's
    Problem of Temporary Intrinsics (1986: 202–5) which leads him to
    argue that entities *perdure*, they are temporally extended but not
    wholly present at any specific point in time, just as a road is not
    wholly present at any specific point in space. This contrasts with the
    conception of persistence known as *endurance*, which requires that
    some core intrinsic feature or essence of an entity continues to exist
    through time in order for that entity to persist.

6.3 Proposed criteria for the distinction

6.3.1 Loneliness

Jaegwon Kim (1982: 59–60) attempted to distinguish intrinsic from extrinsic
properties, drawing upon the intuitive characterizations which have been
sketched above:

> *(Loneliness)*: P is intrinsic   *if and only if*   It is possible that some object b
>   has P although no other contingent object wholly distinct from b exists.

If a property's instantiation by an object does not depend upon the existence
of any other contingently existing objects then, one would think, that property
would count as intrinsic. The specification that we are only concerned with
contingently existing objects is important since it is trivially impossible for an
object to exist independently of necessarily existing objects, should any exist,
even though these intuitively make no difference to whether most contingent
properties can be instantiated or not.

Despite its simplicity and intuitive appeal, Kim's loneliness criterion fails,
as Lewis quickly pointed out (1983b: 198–9). To illustrate this, Lewis defined
the properties of *accompaniment* and *loneliness*: an object is *accompanied* if
and only if it coexists with some wholly distinct contingent object, and *lonely*
if not. Now Kim's definition can be read as saying that a property is intrinsic

if and only if it can be instantiated by a lonely object. However, this definition is insufficient, since having the property of *being lonely* intuitively depends upon there being no other objects; *being lonely* is extrinsic; and yet it satisfies Kim's criterion of being intrinsic. His definition does not divide intrinsic from extrinsic properties in precisely the right way.

6.3.2 Duplicates

Lewis's first attempt at an alternative distinction relies upon the close relationship between particulars having all the same intrinsic properties and being duplicates of each other: if b instantiates intrinsic property P, then its duplicate c will also instantiate P, although b and c can differ with respect to their extrinsic properties. But as Lewis himself admits, the circle of definition is tight: b is a duplicate of c if and only if b exemplifies all the same intrinsic properties as c (1983b: 202). As yet, not much progress has been made.

However, Lewis thinks that we can break into this circle of inter-definition by analysing duplication in terms of shared perfectly natural properties and thereby avoiding recourse to the sharing of intrinsic properties to explain duplication. In Lewis's view, all perfectly natural properties are intrinsic, but not all intrinsic properties need be perfectly natural, since there might be properties shared between duplicates which are compounds of perfectly natural properties and thus are merely natural. For instance, let us say that *having mass* and *having negative charge* are perfectly natural, and thereby intrinsic; this would entail that the conjunctive property of *having mass and having negative charge* was also intrinsic, since both of its conjuncts are, although it would not count as a perfectly natural property.

There are three immediate difficulties with this suggestion: first, one might be concerned about the significant ontological commitment to perfectly natural properties; second and third, one might worry whether such entities are either necessary or sufficient to account for intrinsicality.

Commitment to perfectly natural properties, or to an ontologically elite class of properties like them, is far from universally accepted. It is a useful presupposition to corral a group of sparse properties from the rest, but it is not suited to characterizing infinitely complex worlds, since in such worlds there is no ontologically fundamental base. However, it seems plausible to suppose that intrinsic properties (and duplicates) exist in infinitely complex worlds too, and so it is not obvious that perfectly natural properties are necessary to determine which properties are intrinsic. As was noted in 5.3.1, Lewis is prepared to rule out infinitely complex worlds in order to maintain his ontology; but one might regard this restriction as ad hoc, and consider the need for it to count against the plausibility of perfectly natural properties and their use to define duplication, and intrinsicality, in turn.

In addition, one might wonder whether perfectly natural properties – should there be such entities – are sufficient to account for intrinsicality; that is, whether perfectly natural properties (and boolean combinations of them) will all turn out to be intrinsic. Perhaps some fundamental qualities of the natural world are *extrinsic*, but the existence of such entities would render Lewis's criterion useless. For instance, *mass* might turn out to be an extrinsic property, with its existence requiring the existence of the Higgs Field (Bauer 2011). But if this is the correct characterization of mass, then mass is either not a perfectly natural property, or Lewis's criterion of duplication in terms of perfectly natural properties does not capture the intrinsic-extrinsic distinction.

Finally, even if there is a close relationship between perfectly natural properties and intrinsicality, the current worries suggest that this connection merely holds as a matter of contingent fact, rather than its marking out an essential philosophical association between intrinsicality and perfect naturalness (Yablo 1999: 480).

6.3.3 Independence from loneliness and accompaniment

Rae Langton and David Lewis (1998) suggested an improvement on Kim's loneliness criterion which can bypass Lewis's original complaint that the property of *being lonely* turns out to be intrinsic on Kim's account. They note that being an intrinsic property is *independent* of both accompaniment and loneliness; it simply does not matter to the instantiation of an intrinsic property whether any other entities exist or not.

However, this distinction does not apply to all properties and some qualifications need to be made. First, Langton and Lewis simply stipulate that their criterion does not apply to properties which involve specific objects, such as *being written by David Lewis*, *being Obama*, or *being more populous than Dhaka*. These are sometimes called *haecceistic* properties because they involve a specific entity (this person David Lewis, this person Obama, or this city Dhaka), or *impure* properties to contrast them with *purely* qualitative properties. Since each concerns a specific individual, their instantiation cannot be independent of loneliness and accompaniment, so Langton and Lewis's criterion cannot apply.

Secondly, the criterion must explicitly exclude disjunctive properties. For example, consider the property of *being spherical and lonely or non-spherical and accompanied*: it is independent of loneliness and accompaniment, since a lonely thing may have it or lack it, and an accompanied thing may have it or lack it, but it is not intrinsic. Finally, the negations of disjunctive properties must be disallowed, since if a property is intrinsic, its negation is also intrinsic, and similarly the negations of extrinsic properties are also extrinsic.

But what counts as a disjunctive property? One might be worried that *any* property could be suitably twisted into a disjunction and prohibited by Langton and Lewis's restriction on disjunctive properties, thus rendering the set of intrinsic properties empty. For example, take the properties of *being grue* and *being bleen*; the former is defined as *being green if observed before t, otherwise blue*, while the latter is defined as *being blue if observed before t, otherwise green*. While grue and bleen are themselves disjunctive properties, they can be put into a disjunction in order to make green into a disjunctive property: something is *green* if and only if it is *grue or bleen*. More formally, and more generally, any property F can be expressed disjunctively since any object is *F* if and only if it is *F and G or F and not G*. Langton and Lewis need to rule out such gerrymandered properties in order to retain a core set of properties which are not disjunctive and they do so by recourse to the primitive notion that some properties are more natural than others. Thus, an apparently disjunctive property does not count as genuinely disjunctive if its disjuncts are less natural than the disjunction as a whole; or, conversely, a genuinely disjunctive property is one with disjuncts more natural than it is. This rules out the counterintuitive disjunctive examples, such as *being grue or bleen* above, and provides a criterion of *basic* intrinsic properties, in terms of which all other intrinsic properties can be constructed in terms of conjunctions, disjunctions or arbitrarily complicated truth-functional compounds of basic intrinsic properties.

We can summarize Langton and Lewis's distinction as follows:

(Independence): P is an intrinsic property *if and only if*
    1) P is independent of loneliness and accompaniment
        i.e. (i)   It is possible for a lonely thing to have P
            (ii) It is possible for a lonely thing to lack P
            (iii) It is possible for an accompanied (non-lonely) thing to have P
            (iv) It is possible for an accompanied thing to lack P
    2) P is not a disjunctive property
    3) P is not the negation of a disjunctive property

In contrast to Lewis's original account, the independence criterion does not rely upon a primitive ontological assumption about the mind-independent existence of perfectly natural properties which are themselves intrinsic; all that is needed here is a scale of naturalness, a hierarchy of properties to ensure that not all properties are trivially disjunctive. Moreover, the measure and nature of that naturalness is open: it could be an objective feature of properties, as in the accounts of sparse set-theoretic properties, universals or tropes considered so far; or else naturalness might be determined by the importance of a property relative to our theories or the way we think. Langton and Lewis highlight this flexibility as an advantage of their amended criterion over Lewis's original suggestion (1998: 119–20), although this does not indicate that Lewis himself changed his opinion from his previous view; rather, he recognized that explaining intrinsicality via the role of perfectly natural properties in duplication requires a greater ontological commitment than the independence account and is not as widely acceptable as a result.

One may still have some qualms about the revised account. First, as Langton and Lewis accept, the independence account does not work if there are necessary connections between properties themselves; that is, if either the existence of properties is brought about by necessary laws of nature via the existence of other properties, or properties themselves involve some essential or necessary causal power to bring other properties about. In such cases, fewer properties will be independent of the existence of other ones – perhaps none will be independent as required – and so fewer properties will count as intrinsic. The question of whether there are such necessary connections between instrinsic properties will be considered in Chapters 7 and 8. For the moment, it is sufficient to note that necessary connections between properties are only disastrous for the independence criterion of intrinsicality if it is impossible for situations to obtain which are contrary to the laws of nature. Thus, if the modal strength with which properties are connected is weaker than metaphysical necessity, we can construe the possibility in the independence criterion as metaphysical, and such possibilities can obtain regardless of nomologically necessary connections between properties. Langton and Lewis's view is only inconsistent with a specific account of modality which identifies nomological and metaphysical necessity. Nevertheless, as we will see in 7.4 and 8.4, this univocal conception of necessity turns out to be very plausible for those who consider properties to be essentially causal entities, in which case the independence criterion will have to be abandoned.

A second problem is associated with some properties which are locally intrinsic, but not globally so; for instance, the property of *being such that there is a sphere* (Marshall and Parsons 2001). This existential property is intrinsic according to the independence criterion: it is independent of loneliness, since it can be had by a lonely sphere and by an accompanied one, and lacked by a lonely cube and an accompanied one; it is not disjunctive; nor is it the negation of a disjunctive property. It should turn out to be extrinsic and it does not. Langton and Lewis's response (2001) is to argue that *being such that there is a sphere* is legitimately disjunctive. It is equivalent, they maintain, to *either being a sphere or being a non-sphere accompanied by a sphere*, a property with disjuncts which are more natural than it is, and as such is extrinsic as required.

A third objection to the independence account concerns what Ted Sider (2001) has called *maximal* properties. Consider the property of *being a rock*, for example, typically exemplified by large independent lumps of stone and lacked by other things. While a particular rock (call it Rock 1) exemplifies the property of *being a rock*, it has proper parts which do not have the property of *being a rock* because they are attached to other parts of Rock 1. (c is a proper part of d *if and only if* c is a part of d and c is not identical to d.) For instance, take the proper part of Rock 1 which includes all of Rock 1 except for the 10 cm thick layer of rock over Rock 1's whole surface and let us call this R*1. R*1 does not exemplify the property of being a rock because it is surrounded by the rest of Rock 1. If the outer layer were chipped away, then R*1 would be a rock too. But now it seems that the fact that Rock 1 exemplifies *being a rock* is dependent upon its not being connected to other pieces of rock; for example, by its not being surrounded by a 10 cm layer of rock just as R*1 is surrounded by the rest of Rock 1. *Being a rock* is an extrinsic property, because a particular's having or lacking it depends upon the existence of entities which are not identical to that particular.

However, *being a rock* is intrinsic according to the independence criterion: it is independent of loneliness because there could be a possible world in which only a rock existed, a world in which an accompanied rock existed, a world in which a lonely non-rock existed (a lonely frog, for instance), and a world in which a non-rock was accompanied by other things. The independence criterion puts *being a rock* (and other maximal properties, such as *being a house*) in the wrong category, unless they can be treated as being disjunctive properties. For instance, we could attempt to distinguish *being a rock* into an intrinsic component *being a rock**, and a maximality component such that something is a rock if it is a rock* and not a proper part of another rock*; so, *being a rock* is the same as the negation of *being a non-rock** or *being a rock* that is part of another rock**. To ensure that this is a bona fide disjunctive property, the property of *not being a rock* would have to be less natural than either of the former property's disjuncts, and that does not seem plausible at all. Moreover, it is even less obvious how *being a rock* would count as disjunctive on a non-Lewisian account of naturalness in which natural properties are those which play some special role in our thinking, or are those which are central to our theories, or are picked out by our most important theoretical terms (Taylor 1993; Hirsch 1993). It seems likely that ordinary sortal terms such as 'house' and 'rock' will count as fairly natural in such systems and so we cannot really recast them as disguised disjunctions, or as negations of disjunctive properties, thereby disallowing them as basic intrinsic properties. *Being a rock* and other maximal properties wrongly count as intrinsic on the independence account.

One solution, favoured by Sider, is to return to Lewis's criterion with its narrower conception of naturalness which relies on the objective existence of perfectly natural properties. These mind-independent features of the world serve to rule out *being a rock* and other maximal properties from counting as intrinsic, because they can neither be constructed nor ontologically determined by intrinsic perfectly natural properties alone. However, as mentioned previously, the ontological commitment involved in Lewis's criterion is quite significant, and so alternative criteria for distinguishing intrinsic from extrinsic properties have been sought.

6.3.4 *Grounding*

One suggestion aims to revise the Langton-Lewis account in order to remove the reliance on naturalness entirely. This characterizes intrinisicality in terms of a property's being independent of loneliness and accompaniment and also its being *grounded* by other properties also had by the individual. Like naturalness, grounding is a metaphysical notion and so the account is not purely logical, but its proponents claim that grounding is less opaque and obscure than Lewis's notion of naturalness and thus that their account has greater clarity (Witmer, Butchard and Trogdon 2005). Recall that naturalness was invoked to distinguish between genuinely disjunctive properties, which are extrinsic, from artificial disjunctions, thereby dealing with intuitively extrinsic properties which are independent of loneliness and accompaniment. However, we might be able to achieve the same result by insisting that a property is only intrinsic if it is instantiated *in virtue of* the other properties instantiated by the same individual, when these properties are also independent of loneliness and accompaniment. From the initials of its main proponents, this account has been called the *WBT Grounding* account.

The proposal involves two conditions:

WBT Grounding

A) Property P is had by individual b in *an intrinsic fashion* iff
    (a) P is independent of loneliness and accompaniment;
    (b) If P is instantiated by b at least partially in virtue of other properties of b, each of these is also independent of accompaniment.

B) Property P is *intrinsic* iff any possible individual x which has P has P in an intrinsic fashion.

There are two points to note about this definition: first, that 'having a property in an intrinsic fashion' is a relativized notion: a property F may be instantiated in an intrinsic fashion by some individuals and not by others which nevertheless instantiate it. For instance, the property of *being a cube, or non-spherical and accompanied* is had in an intrinsic fashion by cubes, while it is not had in an intrinsic fashion by non-spherical things which are accompanied by some other contingent individual, when the property is instantiated by an accompanied pentagon (say). Similarly, the property of *being such that a black dog exists* is instantiated in an intrinsic fashion by any black dog and not by other individuals who coexist with black dogs. If this relativization is taken to be problematic, then the proposal is not viable and will have to be abandoned (Hawthorne 2001: 400 n.1; WBT 2005: 333–4). But it is not clear why it should be a problem, since the relativization is similar to the characterization of locally intrinsic properties at the beginning of this chapter. What constraint (B) ensures is that only those properties which are locally intrinsic in all possible individuals which instantiate them count as intrinsic. There does not appear to be a viable objection here.

Second, the definition conflicts with the criterion of property identity based upon necessary coextension because, on the WBT grounding account, necessarily coextensive properties can differ in whether they are intrinsic or not (WBT 2005: 334). For example, consider the intrinsic property Q and the property *being Q and accompanied or Q and lonely*; clearly, any individual which has Q is also either lonely or accompanied, and so also instantiates the property of *being Q and accompanied or Q and lonely*, but while Q is intrinsic, the latter property is not. This may be taken to indicate that the identity criterion for properties should permit finer-grained distinctions between properties – so that *being Q* is distinct from *being Q and accompanied or Q and lonely* – suggesting the plausibility of a hyperintensional criterion. Or, it may be taken to imply that the properties of *being intrinsic* or *being extrinsic* are not directly possessed by properties as we have been assuming, but are instantiated by finer-grained entities such as predicates, meanings, or Fregean concepts instead. In the latter case, whether a property is intrinsic or not depends upon how it is described, upon the predicate used to pick it out. WBT consider this dilemma to be an acceptable consequence of their proposal (2005: 334), but it is not clear that either option is a satisfactory one. There will be more to be said about the relationship between the intrinsic-extrinsic property distinction and property identity in 6.5, since it is not solely a problem for the grounding criterion.

In addition, one might raise some serious objections to the use of grounding in this proposal as it stands. First: what does 'in virtue of' mean? Unless we can adequately explain what the 'in virtue of' locution means in this context, then WBT grounding has not delivered the advantage over the naturalness account which its proponents promised. It is, on the most pessimistic assessment of it, either obscure, or circular, or it does not define 'intrinsic property' as we usually intend such a notion. It would be obscure were our understanding of 'in virtue of' no clearer than our understanding of property naturalness; but WBT reject this complaint, arguing that the 'in virtue of' locution is widespread through a swathe of philosophical discourse, while property naturalness is a specialized notion, invoked as a primitive in Lewis's theory. But this response may be too swift, since the popularity and entrenchment of a phrase in philosophy does not automatically yield trans-parency of meaning, and furthermore, its use might not be unequivocal in all philosophical contexts: just because philosophers use a phrase does not entail that they know what it means and certainly does not imply that they are all describing the same phenomenon. On the other hand, Lewis's notion of naturalness is introduced for the specific task of capturing the relation between a property and more fundamental properties and, although it is a novel term, it may claim to be just as clear as 'in virtue of'. One might worry that the same defence is not available for Lewis and Langton's conception of naturalness, since it is much broader than Lewis's objective notion; but one could respond that each specific version could be made just as clear.

Despite these worries about grounding, I will allow that a suitable definition might be available, although there is not space to evaluate suggestions here.[2] Or else, grounding could be treated as a primitive metaphysical relation and intrinsicality defined in terms of it. So the problem of explicating grounding is not fatal to the WBT account.

However, this difficulty prompts another worry that the use of the 'in virtue of' locution is circular because it presupposes a prior notion of intrinsicality. Although WBT deny that their account is directly circular (2005: 340–7) – that is, that 'in virtue of' cannot be understood without recourse to the notion of intrinsicality – there is a more subtle difficulty which renders their criterion insufficient without the intrusion of intrinsicality in the *definiens*, on the right hand side of the biconditional in (A).'(A) states that for a property P to be instantiated in an intrinsic fashion in a particular, the property P and any property upon which P at least partially depends must be independent of accompaniment. So, if b has P in virtue of Q, Q must be independent of accompaniment. The following objection concerns the status of properties such as Q upon which the property P (whose intrinsicality is at issue) depends, or in which P is grounded.

It is now well known from other examples in this chapter that a property can be independent[3] and yet extrinsic: an existential property such as *being such that there is a bear* can be had by a lonely bear and an accom-panied bear, lacked by a lonely human and an accompanied human; as can disjunctive properties such as *being spherical and lonely or non-spherical and accompanied*. In the latter case, such properties can never be had in an intrinsic fashion because whichever side of the disjunction a particular instantiates, the instantiation of the disjunct is at least partially in virtue of properties which are *not* independent of accompaniment, in this example either *being lonely* or *being accompanied*. In the former case, properties such as *being such that there is a bear* can be instantiated by some individuals in an intrinsic fashion and others in an extrinsic fashion (an implication accepted by WBT 2005: 334). Such properties are prevented from being intrinsic *tout court* because they do not meet condition (B): that is, they are not had in an intrinsic manner by any possible individual which instantiates them. Such a property can legitimately act as a grounding property for locally intrinsic properties in a particular b if it is instantiated by b in an intrinsic fashion, but not if it is instantiated in an extrinsic fashion. However, the WBT grounding account has difficulty distinguishing between these two cases without surreptitiously stipulating that only those grounding properties had in an intrinsic fashion are permitted, and that makes the account circular. For instance, *being such that there is a bear* can legitimately ground (locally) intrinsic properties in bears, but not in non-bears, but condition (A) is not strong enough to enforce this distinction, and attempts to make it so lead around in a circle or into a regress.

For example, what if a locally intrinsic property (let us call it 'Q') is among the ones in virtue of which a property P is instantiated? Condition (A) requires that Q is independent, but this could mean one of two things: either, that Q is independent in general (that is, Q satisfies Langton and Lewis's four conditions of independence), or that individual b instantiates (or could instantiate) Q independently of accompaniment. Let us call these *independence*$_1$ and *independence*$_2$ respectively. As we have seen, independence$_1$ is too weak to prevent Q from being extrinsic: simply being independent of accompaniment in general is insufficient to rule out extrinsic properties, and so condition (A) could be met by a property P which depended upon an extrinsic property Q of b. In such cases, it seems wrong to say that P is had by b in an intrinsic fashion. But if we take the other horn of the dilemma and require that property Q is not just *independent*$_1$, but also instantiated by b independently of accompaniment in this particular instance – that Q is *independent*$_2$ – then we should ask what this more specific independence$_2$ amounts to.

We must resist the temptation to say that for Q to be *independent*$_2$ is for b to instantiate Q in an intrinsic fashion, because that would be circular. So it seems that in saying that Q is independent$_2$ we are saying that b instantiates Q in virtue of how b is; that is, in virtue of other things about it, namely other properties of b (say, those in set **R**). But then a regress threatens, since each of the properties in **R** must also be independent$_2$, properties which could be independent of accompaniment when instantiated by b; this, in turn, can only be ensured by the independence$_2$ of properties upon which members of the set **R** depend when instantiated by b, and so on. The regress will be halted by properties of b which are not instantiated in virtue of any others and are instantiated by b independently of accompaniment. But, why should an arbitrary particular b have such fundamental properties *and* have them in an intrinsic fashion? While Lewis required a primitive presupposition about the existence and intrinsicality of perfectly natural properties, the grounding criterion could halt the regress by assuming that individuals have fundamental properties in an intrinsic fashion. But WBT grounding was intended to avoid such specific ontological commitment.

The attempt to make condition (A) sufficient using independence$_2$ results in either a regress of dependency or a circle because we require grounding properties to be instantiated in an intrinsic fashion themselves (not merely to be independent$_1$). Without a significant ontological assumption about the existence of fundamental (locally) intrinsic properties to halt the regress, (A) and thus the WBT grounding criterion is inadequate to distinguish between intrinsic and extrinsic properties.

Perhaps the WBT account can be amended to give an independent story about how grounding is different when the properties doing the grounding are themselves intrinsic. For instance, Rosen (2010) suggests that we can give a grounding-based account of intrinsicality which utilises the mereological notion of the instantiation of the properties of an individual depending upon the nature of the individual and its parts. The challenge for this account will be to explicate this latter notion without presupposing intrinsicality once again. There is more to say on this issue, but I will leave the matter here.[4]

6.4 Can we define extrinsicality?

One broadly logical, metaphysically neutral version of the intrinsic-extrinsic property distinction attempts to characterize extrinsicality rather than intrinsicality. If these two kinds of properties are mutually exclusive, and between them exhaustively cover the category of properties, then defining extrinsicality will provide a definition of intrinsicality as well. This criterion is a more sophisticated descendant of Kim's original proposal which attempted to draw the distinction by characterizing intrinsic properties as those which are compatible with loneliness. It begins by noting that the instantiation of an extrinsic property by an individual consists in it bearing certain relations to at least one distinct concrete individual. Let us call properties which do this d-relational properties. Properties which are not d-relational will be intrinsic. We can now attempt to understand d-relational properties more precisely.

6.4.1 D-relational properties

Although formulations vary, one can identify three types of d-relational properties (Francescotti 1999; Harris 2010: 467).

Types of d-relational properties

1 P is a (positive) *impure* d-relational property of individual b iff there is a relation R and an individual c such that:
>   (i) b's having P consists in b's bearing R to c;
>   (ii) b is distinct from c.

2 P is a (positive) *pure* existential d-relational property of individual b iff there is a relation R and a class C of individuals such that:
>   (i) b's having P consists in b's bearing R to a member of C;
>   (ii) there is an element of C which is distinct from b.

3 P is a (positive) *universal* d-relational property of individual b iff there is a relation R and a class C of individuals such that:
>   (i) b's having P consists in b's bearing R to every member of C;
>   (ii) *it is possible that* there is an element of C which is distinct from b.

D-relational properties of type (1) include properties such as *being 85 km from Oxford*, which I currently instantiate by bearing a relation to a particular, the city of Oxford, which is distinct from me. By 'distinct' individuals, we mean individuals which are neither identical to each other, nor is one a proper part of the other. (2) includes locally intrinsic properties such as *being frightened by bats* or *being such that there are dogs*, which could be instantiated by a bat or a dog respectively, but would count as extrinsic overall because the particular instantiating the property in such cases would be included as an element in the class C to which it bore the relevant d-relations. Finally, properties of the third type include superlative properties, such as *being the tallest person in the room*; such properties are contingent upon which other people are in the room – I would cease to instantiate that property if someone taller walked in – but we have to allow that there may be no-one else in the room (since I am currently alone, I am instantiating the property of *being the tallest person in the room*).

However, some intuitively extrinsic properties turn out to be intrinsic according to the definitions above. For example, the property of *not being close to a bat* can be instantiated by a lonely object – an object in a world on its own – as long as that object is not a bat. It is not a d-relational property of any of the three types, because it can be instantiated without the object which exemplifies it being related to an individual or a class of individuals, and yet it is intuitively an extrinsic property.

We can rectify this problem if we allow d-relations to hold between individuals in different possible worlds as well as their being intra-world relations. In the case of (3) above, this scenario is already implied, since the class C of individuals to which a particular b instantiating a d-relational property is related already includes elements which are possible with respect to b itself; that is, C includes entities which exist in a distinct possible world from b. If b is d-related to those entities, then d-relations can be inter-world as required. In the case of (1) and (2), we need to add the requirement that the relation between b and the distinct particular or particulars, can be between actual or possible particulars, not simply between actual ones.

6.4.2 D-relations and abstract objects

A more serious difficulty concerns whether an individual should be allowed to be d-related to abstract objects, or only to concrete ones. Either option has unpalatable consequences, and there is disagreement among the proponents of this view about which strategy is best.[5] On one hand, if an individual's bearing a relation to an abstract object (or a class of objects) distinct from itself counts as a d-relation, then measurements turn out to be extrinsic properties on the assumption that numbers are abstract objects. On this view, my desk's instantiating the property of *being 110 cm wide* is an extrinsic property of the desk because its instantiation consists in the desk being related to a number. However, this stance seems slightly peculiar: intuitively, intrinsic properties are all and only those which are shared by duplicates and we would expect that duplicates would share measurements. A desk which is qualitatively identical to mine had better be 110 cm wide, but that means that measurement is intuitively intrinsic contrary to the d-relational account under discussion. Second, such an account might result in *every* property being extrinsic, since the relation of *exemplification* would turn out to be d-relational, and so would be extrinsic, were properties to be abstract objects such as transcendent universals. Given this ontological account of properties, an individual's *being red* would consist in it bearing an exemplification relation to a universal *Red* which is distinct from it, which would rule out *being red* from being intrinsic. (If you are unhappy with colours as intrinsic properties, substitute a property you do consider to be intrinsic and the same conclusion will follow.)

One could try to avoid these two conclusions with an account of number or measurement which avoided abstract objects, and an account of the ontological basis of properties which did not treat properties as abstract objects. But it seems ad hoc to motivate decisions about basic ontology to prop up an account of the distinction between intrinsic and extrinsic properties. On the other hand, one could opt for the opposing position that d-relations cannot hold between an individual and an abstract object or objects; and thus, that a property's being intrinsic cannot be ruled out by its instantiation involving a relation to an abstract object. On this view, both concrete and abstract individuals can exemplify properties, but such properties only count as d-relational when the individual is related to other concrete *particulars*, either possible or actual, and not when it is related to abstract entities. If one adopts this strategy, the problem with measurements is solved and measurement properties are intrinsic again as intuitively required,[6] whether or not numbers are abstract. Furthermore, *exemplification* is intrinsic even in the case of transcendent universals. Third, the properties of abstract objects such as numbers, which numbers instantiate because of their relations with other abstract objects, also turn out to be intrinsic on this view: for instance, the number three *is prime*, or *is less than five,* and these are (arguably) intrinsic properties of the number three, although they are instantiated because of three's relations with other numbers.

But there is a danger that this restriction on the relata of d-relations is too strong since it renders all properties which involve relations to abstract objects intrinsic. For instance, indiscriminately necessary properties such as *being such that 37 exists*, or *coexisting with roundness* will both be intrinsic if roundness and the number 37 are both abstract. Every individual will have countless intrinsic properties due to its relation to every necessarily existing abstract entity, properties which add nothing to its qualitative similarity to or difference from other particulars (since everything has such properties).

However, perhaps one could live with these consequences for two reasons. First, indiscriminately necessary properties are instantiated by every concrete particular, possible and actual, and thus they are necessarily coextensive with each other. If property identity is a matter of necessary coextension, then these properties are numerically identical and we need not consider each object to exemplify uncountably many intrinsic properties. On the other hand, if one opts for a finer-grained account of property identity, one could invoke a version of the distinction between sparse and abundant properties to minimize the ontological implications: properties which are instantiated by every concrete object are not candidates to be among those which pick out natural divisions in the world, since they cannot form the basis of relations of difference between coexisting objects, and so they are only vacuously involved in characterizing relations of resemblance. One might object that this introduces a non-logical aspect into this version of the distinction between intrinsic and extrinsic properties, since it may be based on a notion such as naturalness. However, distinguishing sparse from abundant properties is not an essential part of making the distinction between intrinsic and extrinsic properties in this instance, rather it helps to explain away a counterintuitive result: while we intuitively tend to associate intrinsicality with sparse properties, these are not the only intrinsic properties which there are.

This position on all properties involving relations to abstract objects being intrinsic is reinforced by the fact that those properties which involve relations to abstract objects and which are not indiscriminately necessary do seem to be intrinsic and to mark out resemblance relations between objects. For instance, there is a property which Russell instantiated and Quine lacked of *endorsing transcendent universals*. Such a property involves a relation to abstract objects, but it is not indiscriminately necessary, and it seems plausible to argue that the reason for this is that it marks out genuine differences between the individuals who instantiate it and that it is the kind of property which we would expect duplicates to share. It is not counterintuitive to treat these properties as intrinsic.

6.5 Intrinsicality and property identity

Some of the potential criteria of intrinsicality discussed so far are at odds with the account of property identity based on necessary coextension. This issue was raised in relation to the WBT grounding account (6.3.4) but it also affects the d-relational definition of extrinsicality (6.4) and the Lewisian accounts based on duplication and naturalness (6.3.2, 6.3.3), if these are taken to include identity properties, and indiscriminately necessary properties which every entity has (Eddon 2011).[7]

The problem is that if necessarily coextensive properties are identical, then one property can be both intrinsic and extrinsic according to how it is described, which is obviously impossible. For instance, the properties of *being a bear* and *being a bear and lonely or a bear and accompanied*; or, *being a cat* and *being a cat and coexisting with 37*, are pairs of necessarily coextensive properties but the former is intrinsic, while the latter is not. (On the d-relational account which treats indiscriminately necessary properties as intrinsic (6.4.2), then the latter example does not apply.) If we wish to stick to this version of property identity, we have to alter what intrinsicality itself is a property of; namely that it is a property of the predicates, or concepts which we use to pick properties out. But that seems unsatisfactory, since the uses to which the intrinsic-extrinsic property distinction are put are explicitly metaphysical and not simply observations about language use. For instance, when we say that the book which I threw had the causal efficacy that it did in virtue of its intrinsic properties and not its extrinsic ones, it does seem that this point concerns a second order property (the intrinsicality) of real, objectively existing entities, and that such intrinsicality would not disappear were those entities to be re-described.

If we trust this feeling of dissatisfaction, then it might seem that we are forced to say that the ontology of properties is finer-grained than identification based on necessary coextension can provide and is sensitive to hyperintensional differences (5.2). But do we have to accept this conclusion on the basis of problems defining intrinsicality? And, do we have to accept it for all properties? Perhaps we could exclude the problematic haecceistic and necessary properties and concentrate on defining the distinction for pure qualitative properties, as Lewis and Langton do. We could say that the problem cases are, in the main, silly properties 'that only a metaphysician would ever think of', like *being such that there possibly exists something greater in mass* which is instantiated by every possible individual or *being such that 37 exists* (Marshall and Parsons 2001: 349). Or, slightly less provocatively, we might say that they are properties whose intrinsicality is not important, or just different from the intrinsicality of qualitative properties for which non-conflicting criteria for both property identity and intrinsicality can be provided.[8] It may be no coincidence that disagreements concerning impure, or indiscriminately necessary properties have been some of the main sources of contention between accounts if they require an alternative criterion, or are perhaps not susceptible to being classified as intrinsic or extrinsic at all.

But if we adopt a policy of 'divide and rule' to make the intrinsic-extrinsic property distinction, then this depends upon our being able to make the distinction between qualitative properties and the rest cleanly, and then formulating accounts of intrinsicality to fit the different cases. Perhaps a return to the distinction between sparse and abundant properties is called for, for which a primitive metaphysical distinction will be required. After all, the metaphysical uses of intrinsicality in 6.2 are most relevant to sparse properties, or properties closely ontologically dependent upon them; not to any property for which we can formulate a name. Put in this way, Langton and Lewis's reliance on property naturalness does not seem out of place. Or, alternatively, one might accept a version of the d-relational account which does not need this distinction. In either of these cases, there is a plausible account of intrinsicality which does not conflict with the criterion of property identity based on necessary coextension.[9]

6.6 Is there a single notion of intrinsicality?

Some readers might be concerned at this point about the state of the debate; they might be unsettled about at least two issues. The first underlying philosophical problem is that decisions between competing accounts of intrinsic and extrinsic properties are being made on the basis of claims for which we have very few – if any – clear reasons except for intuitions of the most flimsy sort. For instance, one might consider that an account's classifying indiscriminately necessary properties as intrinsic is sufficient to exclude it (Marshall 2010), that classifying such properties as intrinsic is harmless (Harris 2010), or that any proposed criterion is not intended to apply to such properties (or at least not to a subset of them). While reliance on intuition in metaphysics is unavoidable, there seems to be a world of difference between accepting the intuition that two red objects share something in common – the Moorean fact of qualitative similarity – and our making intuitive judgements about whether indiscriminately necessary properties, such as *being such that 37 exists*, are intrinsic or not.

The second, perhaps related, reason for concern is that there may not be a single way in which to understand the distinction between intrinsic and extrinsic properties. Before Kim's proposal, the intrinsic-extrinsic distinction was not well-examined, and general assumptions tended be made about what makes a property intrinsic or extrinsic, such as its being non-relational or relational,[10] with little examination of the implications which these assumptions had. The more precise formulations of the distinction might capture slightly different aspects of our intuitive idea of intrinsicality. For example, we could isolate three related distinctions, rather than just one:[11]

Table 6.2 Different Conceptions of Intrinsicality

	Distinction	Criterion
1	Duplication-Preserving Properties vs. Properties which do not preserve duplication	P is an intrinsic property iff P never differs between duplicates.
2	Interior vs. Exterior Properties	P is intrinsic iff necessarily, P is a property which any individual x has purely in virtue of how x is, and not in virtue of how anything wholly distinct from x is.
3	Local vs. Non-Local Properties	P is a local property iff necessarily, for any individual x, P is ascribed to x wholly on the basis of how x and its parts are, and not on the basis of how anything distinct from x is.

Initially, it might be difficult to see the differences between these options – which would, in part, serve to explain the conflation of the different notions of intrinsicality in the first place – but the discrepancies between them can be illustrated by the fact that they classify impure or haecceistic properties and indiscriminately necessary properties differently.

For example, indiscriminately necessary properties can never differ between duplicates, because they are instantiated by every possible individual, and so they count as duplication-preserving properties, but they are not local properties, and it is a matter of some disagreement whether they count as interior properties or not. On the latter point, both Sider (1993) and Harris (2010) count them as intrinsic, although their notions of intrinsicality are closer to that of an interior property than to the notion characterized by the dupli-cation preserving account (or, strictly, in Harris's d-relational account, closer to the notion of not-being-an-exterior property). On the other hand, the fact that the instantiation of such properties requires the existence of entities in addition to those which instantiate them, such as numbers or other abstract objects, leads many philosophers to consider these properties to be exterior. Secondly, duplicates do not share haecceistic properties, because these are properties associated with one particular. A particular woman exemplifying the identity property of *being Marie Curie* does not share that property with her duplicate counterparts because they are distinct particulars. Even though they may also be called 'Marie Curie', they are not *this* particular Marie Curie, and their haecceistic properties cannot be shared. On the other hand, it seems intuitively plausible to think that such properties are instantiated in virtue of the nature of the particulars which have them: Marie Curie instan-tiates *being Marie Curie* as a result of the particular that she is, not in virtue of anything outside her. The distinction between Local and Non-Local properties, on the other hand, classifies haecceistic properties as being non-local: while being Marie Curie can correctly be ascribed to a particular on the basis that she is that very person, the fact that Hilary Clinton *cannot* be ascribed the same property of *being Marie Curie* depends upon Marie Curie, as well as Clinton herself. The differences are illustrated on Table 6.3.

Table 6.3 Types of Intrinsicality

	Pure Properties	Indiscriminately Necessary Properties	Haecceistic Properties	Impure Properties
Interior	Intrinsic	Intrinsic (Sider, Harris) *or* Extrinsic (Marshall)	Intrinsic	Extrinsic
Duplication-Preserving	Intrinsic	Intrinsic	Extrinsic	Extrinsic
Local	Intrinsic	Extrinsic	Extrinsic	Extrinsic

6.6.1 Which distinction is the best?

Is one conception of intrinsicality more important than the others? If we take the view that indiscriminately necessary properties are intrinsic, then Duplication-Preserving properties are a proper subset of interior properties, namely those properties which are purely qualitative interior properties. Thus, perhaps, one could make a case for the Interior-Exterior distinction on the basis that it is more comprehensive, being applicable to more properties than the duplication-preserving notion. The odd one out here is the distinction between Local and Non-Local properties because it is characterized in terms of what the *ascription* of a property depends upon, rather than what instantiating a property depends upon. Because of this, we have to consider the individuals which do not instantiate a property Q as well as those which do: do non-Q-instantiating individuals require the existence of other entities outside themselves in order to *be not Q*? If so, Q is non-local even though the entities which do instantiate Q do so as a matter of their own internal nature.

However, if we examine the philosophical uses of intrinsicality, it is not clear that one notion is primary. For example, when we are interested in properties supervening on others, it is commonly maintained that supervenient properties cannot differ between individuals with the same 'base' properties; individuals are duplicates for supervenience purposes if they instantiate the same set of properties upon which the supervenient properties depend. In that case, we had better not include haecceistic properties in the base set, or else no individual will have a duplicate and every supervenience thesis would be trivially true. For supervenience, the Duplication-Preserving characterization of intrinsic properties is the most important one.

However, the distinction between real change from merely Cambridge change turns upon the internal changes in properties which an individual undergoes, rather than those which depend upon other individuals, and so the distinction between Interior and Exterior properties is important, rather than simply those properties which are preserved between duplicates. A change in haecceistic properties can nevertheless count as a real change and this cannot be accounted for by Duplication-Preserving properties. For instance, if we take an individual entity b with a part c, and replace c with an exact duplicate d, then there has been a real change in b although there has been no change in the Duplication-Preserving properties of b. Interior properties are required to characterize real change.

6.7 Conclusions

After investigating some of the main attempts to characterize the distinction between intrinsic and extrinsic properties, based on loneliness, duplication and perfect naturalness, naturalness, grounding and d-relations, we can tentatively conclude the following points. What counts as the most plausible way to draw the distinction depends in part upon the uses for which the distinction is to be put and whether we are content to formulate a distinction which is applicable to a subset of properties rather than all properties. Perhaps the most coherent univocal criterion is that formulated in terms of d-relations, which defines intrinsicality in terms of extrinsicality. Only on the d-relational view does there seem to be a plausible univocal account for abundant properties which is consistent with retaining the identity of properties in terms of necessary coextension.

Although a univocal distinction would be philosophically neater, there is certainly a good argument for accepting Langton and Lewis's restriction to qualitative properties, in order to characterize what counts as a duplicate in accounts of supervenience, externalism and internalism and for assessing intrinsic value. The occasional requirement to consider haecceistic properties – in the assessment of real change, for example – does not preclude these properties being treated differently from purely qualitative properties and these differences may also be important when we consider questions about persistence. On the other hand, the intrinsicality or extrinsicality of indiscriminately necessary properties, upon which philosophical opinions are split, seems devoid of important philosophical implications. Perhaps we simply have too few clear intuitions about the nature of abstract objects to agree upon which side of the distinction they should fall.

FURTHER READING

Lewis 1983b; Langton and Lewis 1998; Witmer, Butchard and Trogdon 2005; Harris 2010.
Some strategies not considered here: Vallentyne 1997; Weatherson 2001.
For objections: Yablo 1999; Marshall 2009, 2012; Sider 2001.

Suggested Questions
1   Why is the distinction between intrinsic and extrinsic properties important?
2   Is there a plausible way to draw the distinction between intrinsic and extrinsic properties which does not involve a substantial metaphysical assumption?
3   Is the WBT grounding criterion circular?

4   Assess the suggestion that a plausible criterion of intrinsicality must apply to all properties, rather than just a subset of them.

5   Does it matter whether indiscriminately necessary properties are intrinsic?

6   Is there one distinction between intrinsic and extrinsic properties, or several?

7   What are the implications if a proposed account of intrinsicality conflicts with an account of the identity of properties?

Notes

1   Real changes are a subset of Cambridge changes, with *merely* Cambridge changes concerning only extrinsic properties (Geach 1969: 71–2).

2   See Trogdon 2013, Schaffer 2009, Bennett (forthcoming) who consider grounding to be univocal, and Hofweber 2009 and Fine 2012 who think that there are different kinds of grounding relations.

3   For brevity, I will abbreviate 'independent of loneliness and accompaniment' to 'independent of accompaniment' or just 'independent'.

4   See Schaffer 2010, Trogdon 2009, Allen (MS).

5   Francescotti (1999) permits d-relations to relate to abstract objects, while Harris disagrees (2010: 471).

6   One might consider measurements to be extrinsic properties because they involve the relation that an object bears to certain units or to a measurement system, but it seems intuitively correct to say that measurements should not turn out to be extrinsic simply because the numbers used to count the units are abstract objects.

7   Langton and Lewis (1998: 118) exclude properties which are impure or haecceistic.

8   Strictly speaking, Langton and Lewis could only deal with the property pairs *being Q* and *being Q and lonely or Q and accompanied*, where Q is intrinsic, if the latter property does *not* count as a disjunctive property. The d-relational account would have to find alternative reasons to exclude this.

9   See Eddon (2011) who is pessimistic about the divide and rule strategy.

10   I will not consider the relational/non-relational version of the distinction because it seems clear that some relational properties are intrinsic since they relate a particular to its parts, or its parts to other parts of it. Intrinsic relational properties include *having more bedrooms than bathrooms* (of a house), *having a heart and kidneys* (of a living creature) and so on.

11   Inevitably, there is disagreement about the ways in which the intrinsic-extrinsic distinction can be understood. See Harris 2010 for an alternative to the options here, which are based upon Weatherson and Marshall 2014.

7

Properties and their causal role: Categorical and dispositional properties

The distinction between categorical and dispositional properties is explicated and some empirical and metaphysical reservations about dispositions are discussed. Arguments that properties have their causal roles necessarily are considered, concentrating upon intuitions about modality and epistemic problems about pure qualities, then counterarguments that metaphysics requires categorical properties are discussed. The dispositionalist is found to have two lines of response to these objections, both of which require metaphysical concessions, but which might be acceptable given the utility of the theory as a parsimonious explanation of modality and laws of nature. The plausibility of an ontology of pure causal powers is briefly considered, including regress arguments, and the conceptual difficulty of causality or potentiality grounding being. Finally, two conciliatory views are investigated which seek to resolve the categorical-dispositional dispute by weakening the distinction between such properties, albeit in favour of versions of dispositionalism.

What is *mass*? There are two contrasting ways to respond to this question. *Mass*, someone might say, is a feature which is essential to everything made of matter, the amount of mass is a measure of how much matter a particular is made of. On the other hand, one might say that *mass* is the property which determines how resistant a particular is to being accelerated when a force is applied, it determines the strength of its gravitational attraction to other massive particulars and how much energy is produced on its destruction. On the former conception, the property of *mass* is something which makes particulars what they are, which endows them with a specific qualitative feature, while on the latter, what is essential to the property is what it can do.

Until now, more emphasis has been placed upon how properties make things the way that they are than upon what properties can do. But, for many philosophers, the relationship between properties and what they cause is a very close one: the causal role of a property may be a necessary feature of it, perhaps essential to its nature;[1] or the causal role of a property may be sufficient to individuate it from other properties, as we saw in 5.4. Taken together, these conditions make properties inherently causal entities, their nature entirely determined by their causal role, and so one might want to identify properties with causal powers or dispositions to behave in a certain way. On the other hand, one might think that this causal conception of properties is empirically mysterious because it endows properties with causal potential that we cannot empirically discover, or that it is metaphysically restrictive because it places too many constraints upon the way the world could be.

This chapter will investigate whether the connection between properties and their causal role is necessary or contingent, while Chapter 8 will explore the implications which this debate has for the relationship between properties and other phenomena, such as causes and effects, laws of nature, and modality.

7.1 Categorical and dispositional properties

We can conceive of properties as qualitative, categorical entities which determine the way things are, or as entities which determine what particulars can do. Of course, as was noted in 5.4, it is plausible to think that some properties, such as mathematical or logical properties, are not causally efficacious. If this is the case, then these families of properties are excluded from the present discussion.[2] However, given that some properties do have a causal role, we can then ask whether they have their respective causal roles necessarily, in all possible worlds.

If the causal role C of a property P is necessary to it, then there are two further options, according to whether having C is sufficient for P being the property it is or not. If having C is both necessary and sufficient for being P, then properties such as P can be identified with their respective causal roles; the essence of such properties is their causal power. On the other hand, if having causal role C is necessary but not sufficient for being P, then another property Q might have C too. In this case, P and Q would differ in virtue of their non-causal aspects, their qualitative nature or quiddity, and the nature of a property would not be exhausted by its causal power.

In what follows, I will call properties which have their causal role as a matter of necessity 'dispositional properties' and those which do not 'categorical properties'. As just noted, even if properties do have their causal roles necessarily, that does not entail that their nature is exhausted by what they can cause. So for now I will leave it open whether any dispositional properties also have an intrinsic qualitative nature or categorical aspect to them. Since I am interested in whether properties have their causal roles necessarily, I will classify such properties on the dispositional side of the distinction for now.

I will restrict my interest in the 'causal role' of a property to what that property can cause – otherwise known as its causal *power* – rather than the potentially wider understanding of 'causal role' which takes into account a property's causal relations to other properties, including both a property's causes and its effects. In the literature on dispositional properties, the terms 'causal role' and 'causal power' are often used interchangeably. However, as I mentioned in 5.4, the conflation of these terms is only harmless given the assumption that properties never differ purely on the basis of their causal origin; that is, there are no distinct properties P and Q which differ only in their respective causes. In what follows, I will assume that it is fairly harmless to overlook these kinds of properties – if they exist – and treat causal 'role' and 'power' synonymously, on the understanding that there may be circumstances in which we want to prise these concepts apart.

Categorical properties are qualitative entities: each categorical property has an intrinsic qualitative nature, sometimes called a *quiddity*. Although these properties might have causal powers, they have them contingently: their causal roles can alter in different possible situations, governed by the laws of nature in each world. Because any such causal role need not be stable across possible situations, the transworld identity of a categorical property is *prima facie* determined entirely by its intrinsic qualitative nature.[3]

One might hold the view that all properties which have a causal role do so necessarily and that there are no categorical properties; or that there are categorical properties alongside dispositional ones, but that the former are causally inert. These views, I will classify as versions of *dispositionalism*, while I will call the thesis that properties have their causal roles contingently *contingentism*. In addition, if one thinks that prima facie both categorical and dispositional properties exist, then one might think that the latter are ontologically dependent on the former, the former on the latter, or that the ontology is dualistic with neither kind of properties reducible to, or dependent upon, the other. Finally, one might take the view that the distinction between categorical and dispositional properties is less philosophically significant than we have supposed, either because the apparently mutually exclusive kinds of properties are not so determinately divided, or because categorical and dispositional are different aspects of the same entities.

7.1.1 *What are dispositions?*

The conception of properties as being inherently causal entities is ancient, but it largely fell out of favour from the Renaissance onwards, when the mechanistic worldview of logically independent properties governed by laws became prevalent. Only recently has this focus changed.[4]

One of the concerns which fuels mistrust of essentially dispositional properties is that their nature is not determined by what they are, nor even by what they do, but by what they *could* do. Unlike categorical properties such as *shape* (say, for the sake of argument) which determine the qualitative nature of a thing, or a way that it is, dispositional properties are present even while they are not being manifested. There are paradigm cases of such dispositional properties in ordinary language: a glass window is fragile – it would break if it were struck – whether or not it is ever struck; a sugar cube has the property of *solubility* even if it never comes into contact with water. Such properties are known as *dispositions*, since they characterize the disposition which an entity has to behave in a certain way if the circumstances are right, or if the right kind of trigger occurs, even though those circumstances may never happen; they are, as Martin puts it, 'ready to go' (2008: 50). Such dispositional properties are clearly causal, but this conception of causation is ontologically richer than that which can be captured in terms of 'Humean' constant conjunctions, in which effects of similar kinds simply follow similar kinds of causes. Dispositions are more than just what they do, since they include a modal aspect which determines what they, or what the particular which has them, *would* do, and perhaps *must* do, in the right circumstances.

The nature of such properties presents two main problems: one metaphysical and the other epistemic. The first, metaphysical argument against them is presented by Armstrong who accuses the supporter of such properties of supporting a form of Meinongianism: dispositional properties are directed towards, or have the potential for, a range of effects, many of which will never occur in the actual world, they are purely potential (Armstrong 1997: 79). The solubility of the sugar cube which is always dry is never actually manifested, so the dispositional property which it has is directed towards a non-existent effect. How many other properties does each particular have which are never actually manifested? To regard such potential outcomes as being part of what there is is to hopelessly over-populate the ontology with non-existent entities (just as Meinong over-populated the universe with particulars which do not actually exist), and to reify the properties which would bring about these never-manifested effects is similarly profligate. Dispositional properties offend against Ockham's razor and are metaphysically mysterious as well.

Secondly, for those of an empiricist persuasion, the existence of natural necessity which makes entities behave in a certain way is difficult to accept as a fundamental feature of the natural world, since it is not clear how we could ever experience or test for such a phenomenon. When an acorn grows into an oak tree, or like magnetic poles repel each other, we observe the same phenomena whether or not these occurrences *must* happen in the way that they do, or if they just *happen* to do so; natural necessity is not open to observation. Thus, there has been a tendency to treat dispositions reductively; that is, to attempt to give an account of dispositions in terms of non-dispositional properties, or to eliminate them from the ontology entirely. This move is either achieved by translating talk of dispositions into non-dispositional language,[5] or by giving an account of how dispositional properties are ontologically dependent upon categorical ones. These approaches seem especially plausible if the examples of dispositions are restricted to a narrow range of cases, such as *fragility* and *solubility*, which might be regarded as less-than-fundamental properties, properties which might arise in some way from the fundamental micro-structural properties of the entities which display them.

However, if we relax the epistemically strict attitude towards ontology which the empiricists and logical positivists endorsed, there are some good reasons for rehabilitating dispositional properties, or causal powers, as entities which cannot be reduced to other kinds of properties. The first is that dispositional properties may be more pervasive and widespread than the limited examples above suggest: physical properties, such as *charge*, *mass*, or *force*, can be treated as dispositional and are characterized in terms of what they do and what they could do; and, on closer inspection, it seems that we determine the presence of many properties by the effect which they have upon us, or upon other things. Properties such as colours, which have hitherto been treated as examples of paradigmatically categorical properties, are known to us because they have certain effects (depending upon circumstances). Dispositional properties are not in such a poor empirical situation in comparison with categorical ones after all.

Second, recall that the prospects of providing suitable identity and individuation criteria for properties were considerably improved by taking the relationship between properties and causal role seriously (5.4). For instance, could a property such as negative charge maintain its identity and yet change its causal role entirely? One might be inclined to say in such circumstances that a different property was present, not negative charge at all. If a property is what it is in virtue of what it does, then the role it plays in causation will serve to identify and individuate it. Less strongly, if properties have their causal powers necessarily, then we are at least assured that properties with different causal roles are distinct.

Third, dispositional properties may play a useful explanatory role as the ontological basis of modality. Because properties on this view are irreducibly modal entities – that is, by their natures they determine what could happen and what must happen, not simply what does happen – then this primitive natural necessity might be sufficient to determine the range of possible situations which there could be. Thus, dispositional properties might work as truth-makers for counterfactual and subjunctive statements. If successful, this would permit an actualist account of modality – one which relies only upon the entities existing in the actual world – to determine what could or what must be the case. A fourth advantage, related to these modal considerations, is that dispositional properties might also ground causal and structural laws of nature, minimizing the need for additional categories of entities to do this work and thus providing a parsimonious and unified ontological picture. These suggestions will be evaluated more thoroughly in Chapter 8.

7.2 Is the causal role of properties necessary or contingent?

7.2.1 Arguments for dispositionalism 1: Metaphysics

What reasons are there for thinking that properties which have a causal role do so necessarily? Two arguments can be developed on the basis of intuitions that causal interactions have some modal force, and that causal statements and laws of nature sustain counterfactuals and subjunctive claims (Schaffer 2005: 6). For instance, iron melts at 1538°C and boils at 2862°C degrees and there is a sense in which iron *must* do so at those temperatures: *any* piece of iron would change state were those temperatures to be applied; and, in an environment of 1538°C, the iron present would not have become liquid had the temperature not been raised that high. There is an intuitive feeling about causation that these interactions must happen in the way that they do, as long as nothing else in the specific circumstances prevents them.

Such intuitions can be marshalled into arguments for the view that properties have their causal powers necessarily:

(1) If properties have their causal powers contingently, then it is possible that like charges attract.

(2) Like charges must repel.

(3) So: Properties do not have their causal powers contingently.

Secondly, the fact that laws of nature sustain counterfactuals requires that properties which act as their truth-makers behave in causally the same way in close possible worlds:

(4) If the relationship between properties and their causal powers is contingent, then nothing guarantees that like charges repel in any possible world.

(5) Like charges must repel in close possible worlds.

(6) So: Properties do not have their causal powers contingently.

The intuitive modal force of causality appears to guarantee that if properties have causal powers at all, they have their causal powers necessarily.

However, a closer inspection of these modal arguments indicates that they are not valid, since they beg the question against contingentism by presupposing a univocal conception of necessity which the contingentist is unlikely to endorse. In the first argument, the problem lies in the fact that in premise (2) 'like charges must repel' only counts as the negation of 'it is possible that like charges attract' in (1) if the modal strength is the same in both sentences. But it is plausible to treat necessity as having different strengths: one might accept that it is metaphysically possible that like charges do not repel each other – that is, that there are possible situations in which two electrons would move towards each other, for instance – while maintaining that they do in all the possible situations which are close to the actual one. In addition to metaphysical necessity and possibility, one can argue that there is also a more restricted form of natural or nomological necessity which serves to partition possible worlds or situations into those which share natural laws. This restricted conception of necessity would underwrite our causal intuitions, while the broader conception of metaphysical necessity and possibility permits the existence of a range of possibilities unrestricted by the laws of nature of the actual world.[6]

Once different strengths of necessity are distinguished, the first argument is obviously invalid, since the metaphysical possibility that like charges do not repel is consistent with the nomological necessity that they do repel. Furthermore, this distinction also renders the second argument invalid, since nomological necessity will ensure that like charges repel in close possible worlds, even though it is metaphysically possible that they do not. Why, though, would one think that such nomological necessity will cover the relevant close possible worlds? The reason for this is purely definitional: part of what it means for a world to be 'close' to the actual one is that the laws of nature are the same as those in the actual world and so, in close possible worlds, the repulsion of like charges is assured (Lewis 1979).

The metaphysical arguments for treating properties as having their causal roles necessarily are not compelling, since both beg the question against contingentism. Furthermore, they are only valid on the dispositionalist's own account if she maintains that there is only one strength of necessity; that is, that what is metaphysically possible determines the same range of worlds as what is possible according to the laws of nature. In the opposing camp, the contingentist needs to give plausible account of restricted nomological necessity in order to make his charge of invalidity stick (Fine 2002; Wilson 2013). I will consider whether it is plausible to regard the laws of nature as operating with metaphysical necessity later; but as yet, neither side is winning the debate.

7.2.2 Arguments for dispositionalism 2: Epistemology

The second set of arguments for dispositionalism focuses upon the untoward epistemic consequences of contingentism, in particular, upon the isolated epistemic situation that contingentism puts us into with respect to which properties there are. Recall that if the relation between a property and what it can do is contingent, then this causal role can vary between possible worlds (governed by which laws of nature there are in those worlds). Properties on this view are essentially qualitative since they determine the way something is, without this including what the property can actually do. Instances of *mass* in the actual world attract each other gravitationally and accelerate in proportion to the force applied to them (at least in frictionless environments), but in another possible world the mass of my desk could make it taste of chocolate and maintain the temperature of the room at a steady −6°C. The range of causal variation, it is argued, makes these pure qualities, or quiddities, epistemically elusive: it is hard to see how the *mass* in my example is *mass* as we understand it at all.

Quiddity can be compared to the notion of haecceity or bare this-ness which is that which makes a particular the individual that it is in isolation from all the qualities that it has. Like quiddity, examples where haecceity matters are conceptually difficult to get to grips with: if the properties of a particular can entirely change and yet the particular retains its individuality and transworld identity, then it is possible that my desk could have been a wombat, or a quasar; and it is hard to see what the transworld identity of particulars amounts to, except that somehow some bare particularity is retained across possible situations by the desk-wombat-quasar thing. Likewise, a quiddity retains its transworld identity across changes of causal role, whatever a particular property is is in no way essentially related to what it does or what the laws of nature can direct it to do (because laws of nature can direct it to do anything); property identity must be a brute qualitative matter.

The epistemic arguments for dispositionalism generate sceptical situations which would arise if the qualitative nature of a property were permitted to be completely divorced from what it can do, and thus from the effects it can have upon us (Shoemaker 1980: 215, 1998: 65–6; Black 2000; Bird 2007; Lewis 2009). Such arguments generally employ one of three strategies to manipulate the properties of worlds in relation to their causal powers, and then make a sceptical claim that we cannot tell the different situations apart. These are known as *Duplication, Permutation,* and *Replacement*, as follows:

> *Duplication*: it is possible for there to be two objects a and b in the same world, which have exactly the same effects upon us and everything else, although they have distinct categorical properties (Pa and ¬Pb is true, say, and ¬Qa and Qb). P and Q have the same causal role.

> *Replacement*: although P plays causal role C in the actual world, an alien property R plays role C in another possible world; the role played by *mass* in the actual world could be played by *schmass* in other worlds.

> *Permutation*: Categorical properties P and Q might globally swap causal roles in another possible world w2; the categorical property which has the *charge* role in the actual world might have the *mass* role in w2, while the property with the *mass* role has the *charge* role in w2.

Contingentism inevitably leads to scepticism, it is argued. If such situations are permitted, we have no way of ascertaining which categorical properties a particular actually has, since (by the Duplication Argument) any effect that a specific particular could have on ourselves or upon other objects in virtue of the properties which it has could equally well be brought about by other properties, by *any* other properties in fact, since any categorical property could have any causal role. Not only that, but because we do not know which possible world we're in – whether we're in the *mass* world or the *schmass* world, for instance – we cannot use scientific investigation to tell us which categorical properties there are. If one refuses to accept that we are in such sceptical scenarios with respect to properties – that is, if one believes that we can know within a reasonable degree of determinacy which properties there are – then contingentism must be false. The causal roles of properties must be essential to them.

Note that if we simply say that a dispositional property has its causal power necessarily, then the scenarios described by the Duplication Argument and the Replacement Argument would also apply to dispositional properties. Nothing rules out two dispositional properties having an identical causal role, as long as those properties have the same role whenever they are instantiated. Thus, these arguments require the stronger claim about the relationship between a property and its causal role, which is often implicitly assumed in such discussions; namely, that property P's having a specific causal power C is necessary and sufficient to be P (i.e. that any property P has its causal power C necessarily *and* causal power C necessarily belongs to P). This necessary one-to-one correlation between properties and their causal roles makes causal role both necessary and sufficient to identify and individuate properties; one might say that it is the essence of the property to have the disposition or the causal role that it does, or that the causal power is constitutive of the property, or identical to it. If this holds of dispositional properties, then Duplication, Permutation and Replacement could not apply to them.

7.3 The contingentist response: Metaphysics needs categorical properties!

The contingentist could try some traditional anti-sceptical responses to alleviate the epistemic problems raised about quiddities, and I will return to consider this strategy in Chapter 9. But scepticism alone cannot force the rejection of a category of entities from the ontology: one can accept the sceptical conclusion and yet maintain that properties have their causal roles contingently (Lewis 2009). Sceptical problems notwithstanding, we might have good reason to accept the existence of properties which are able to behave differently in different possible situations, on the grounds that they are required for several important metaphysical tasks and that disposition- alists are unable to provide a suitable alternative. I will consider four of the primary metaphysical arguments for contingentism.

Recombination/Combinatorialism: There is an attractive account of modality which maintains that different possible situations or worlds are grounded in the recombination of properties (Armstrong 1989b). (This combinatorial view is compatible with the claim that all such properties are actually instantiated, or with the modal realist claim that there are also properties which are possible but not actual, and so the domain of possible situations will vary according to which of these is adopted.) However, this combinatorial account of modality does not sit well with properties having their causal powers necessarily, since the latter view seems to place undue restrictions upon which combinations of properties could occur, a conclusion which can be reached by the following argument:

(1') If the relationship between properties and their causal powers is necessary, then some combinations of *mass* and *acceleration* would be ruled out.

(2') Any combination of *mass* and *acceleration* is possible.

(3') So: The relationship between properties and their causal powers is contingent.

Conceivability. In the philosophy of mind, the conceivability of a zombie – a being which is functionally or physically equivalent to a conscious human subject and yet entirely lacks phenomenal conscious experience – is taken to establish the possibility of zombies (Chalmers 1995). Analogously, the conceivability of properties which behave in a way that is contrary to their actual causal role, such as like charges failing to repel, is employed to establish the possibility of such properties:

(4') It is conceivable that like charges attract.

(5') If the relationship between properties and their causal powers is necessary, then it is inconceivable that like charges attract.

(6') So: The relationship between properties and their causal powers is contingent.

Propositions and Meaning. We can identify a proposition with the set of all possible worlds in which it is true: P = {w: P is true at w}. Although like charges do not actually attract, there is a meaningful proposition that they do attract, a proposition which could be wrongly believed by someone, for instance, and thus there should be a way (modally speaking) for the content of this proposition to arise (Schaffer 2005). However, if properties are dispositional, there is no possible world in which a property such as *charge* behaves in a way which is contrary to its actual causal role; the set of worlds associated with the proposition 'Like charges attract' would be empty and so the proposition would have no meaning (or else, all such propositions would have the same meaning, although that is not a very plausible suggestion).

(7') If the relationship between properties and their causal powers is necessary, then there is no contentful proposition that 'Like charges attract'.

(8') The proposition that 'Like charges attract' does not lack content.

(9') So the relationship between properties and their causal powers is contingent.

Counterfactuals: In order to give an account of the truth conditions of counter-factual claims, it is argued that we need worlds in which properties are able to have different causal roles from those which they actually have, so that properties can act in a way that is contrary to the actual laws of nature. For example, 'If like charges attracted, then Coulomb's Law would be false', or 'If penicillin killed viruses, it could be used to cure the common cold', are both statements which we think could be true were the causal interactions of the world slightly different from those of the actual one. However, we do not just think that such claims are trivially true because their antecedent is false,[7] but because if their antecedent were true, the consequent would be true as well. In order to provide truth-makers for such counterfactuals, Lewis's account of counterfactuals (1973: 75–7) uses the notion of a 'miracle' – a slight violation of the actual laws – to shift like charges into proximity with each other for example, or to extend or alter the antibiotic effect of penicillin to make it an effective anti-viral medicine too. These scenarios are impossible if the causal role of a property is necessary to it, and so an alternative account of the semantics of counterfactuals is required.

Each of these arguments recommends contingentism, implicitly pointing towards some philosophically unpalatable limitations of treating the causal role of a property as necessary to it. The dispositionalist needs responses, some of which will be suggested in the next section. But, despite these, the contingentist argues that dispositionalism still places undue limitations upon what is possible and results in a semantics for counterfactuals which is unduly restrictive.

7.4 Metaphysics without quiddities

The preceding section gave four reasons why metaphysics needs a conception of properties which does not tie them to their causal roles necessarily: recombination; conceivability; to provide content for false propositions; and to provide truth-makers for counterfactuals. The general problem for disposition-alism is that the range of what is possible is unduly constrained if the causal roles of properties determine what could happen; dispositional properties fix, in some sense, what the laws of nature are. In some contexts, this might be regarded as an advantage of dispositionalism, but it disallows possibilities where the properties remain the same as those in the actual world and yet the laws are different, and also those worlds in which the laws are the same as in the actual world and the properties are different.

The dispositionalist's choice of responses depends upon whether she thinks that the properties which exist are restricted to those of the actual world, or if purely possible, alien properties exist which are never instantiated in the actual world. As such, there are broadly speaking two strategies available. Both options involve a certain amount of hard-headedness in the face of the contingentist opposition, although they differ in how, or even whether, they explain away the apparently compelling reasons to think that properties are contingently related to their causal roles.

7.4.1 Biting the bullet: Dispositionalism and actualism

First I will consider the responses of the dispositionalist who wants to restrict her stock of properties to those instantiated in the actual world, although such responses could also be appropriated by someone who was prepared to postulate possible entities too. The first three of the four arguments are dealt with by denying the contingentist's claims: first, properties such as *mass* and *acceleration cannot* recombine in any and all ways; second, if we can conceive that like charges could attract, we are mistaken in how we understand 'charge' and thus no possibility is revealed (or, perhaps, we cannot genuinely conceive of such a possibility at all); and, third, there is no such contentful proposition such that 'Like charges attract' either.

In the first case, recombination is disallowed on the basis that the possibility of endless recombination can only apply to distinct entities, but the properties involved – in this case, *mass* and *acceleration* – are not distinct existences. Because *mass* and *acceleration* are related causally in a lawlike way, their causal powers are not independent of each other and so this will restrict how they can interact and recombine. The same will go for many other properties, and so the limits on recombination will be determined by the range of possibilities determined by the laws of nature which are themselves determined by the causal powers of the properties there are.

Second, we can either deny that we can genuinely conceive of situations in which actual properties behave in ways which violate the laws of nature, or else we can break the link between conceivability and possibility. The latter involves allowing that we can conceive of like charges attracting while denying that this presents a genuine possibility because our conception is based upon a misunderstanding of what charge or attraction is. There may be those who are worried about the link between conceivability and possibility being broken in this way: after all, it seems plausible to allow that anything we can conceive of should count as possible. However, the dispositionalist can respond that the conceivability account of the epistemology of modality is too permissive once we start to consider metaphysical possibilities as being determined by the natures of entities such as properties. We might be able to conceive of logical possibilities (i.e. situations which are not contradictory) which are nevertheless metaphysically impossible, and the contingentist would be begging the question against the dispositionalist's account of modality if he were to count this failure to permit unrestricted possibilities towards the failure of dispositionalism.

Third, the dispositionalist might either deny that there is a contentful proposition such that 'like charges attract' – that is, she can deny premise (8') – or else she could challenge the characterization of a proposition presupposed by the argument, as the set of all possible worlds in which it is true. The latter strategy would be supported by the thought that it is legitimate for someone to believe that like charges attract, even though this is a necessarily false proposition if like charges necessarily repel. Analogously, one might believe a necessarily false proposition in mathematics: for instance, although the Poincaré Conjecture has been proved to be true, one might (in ignorance of the proof) still believe it to be false, and in doing so would believe a necessarily false proposition; similarly, one might believe that Fermat's last theorem is false; and so on. But these propositions are not compatible with identifying propositions with sets of possible worlds. We do not simply want to identify them with the empty set – after all, believing that the Poincaré conjecture is false is a completely different matter (one would reasonably assume) to believing that like charges might attract – and so, if such propositions are to have content, we need a finer-grained account of propositions than that based on the distribution of true propositions over the possible worlds there are. As is the case with the individuation of properties (5.2), the requirement for ultra-fine-grained individuation to respect semantic distinctions may prompt a move from identifying propositions on the basis of their necessary coextension to a hyperintensional account.

Alternatively, the dispositionalist could retain the account of propositions and take the first horn of the dilemma, simply accepting that there is no contentful proposition that 'like charges attract'. But does that response entail that it is impossible to believe that like charges attract? Perhaps she does not have to accept this implication: first, one might suggest that someone could believe propositions which are ultimately incoherent, although grammatically correct; or, secondly, one might suggest that what is actually believed is a different and yet related proposition about an alien property such as *schmarge*. The former option is the only one available to someone who wants to be an actualist – that is, someone who restricts properties to those which are actually instantiated and does not want to commit to an ontology of alien properties – and such a move may provoke objections about whether this mental state genuinely counts as believing at all. The basis of this objection would be that belief is defined as having an attitude towards a proposition, such that 'S believes that P' can be broken down as 'S has a believing attitude towards proposition P'. But for this gloss to work, P must be a proposition, in particular P must be declarative or truth-apt, and an incoherent proposition does not have the requisite content, it is simply not something one can have a *believing* attitude towards. However, such an objection need not be fatal; perhaps in such cases the dispositionalist could adjust her position and accept that the subject is not believing as such, but that he is in another related psychological state which does allow one to regard incoherent propositions as if they were true.[8]

7.4.2 The retreat from actualism: Schmarge to the rescue?

Dispositionalists who are prepared to accept the existence of alien properties[9] appear to be able to do a better job at explaining away the apparent need for contingentism in metaphysics. They can do so because, in addition to the hard-headed response that restricts which possibilities there are, they can also explain away the appearance of such possibilities and provide analogous alternatives. To do so, they can explain why one might be deluded into thinking that properties and their causal roles could come apart: when we appear to consider a situation in which like charges attract, and the implications that such a situation may have, we are *misdescribing* an alien situation concerning the behaviour of like instances of an alien property very much like charge, *schmarge* for example. The possibility that like charges attract is only apparent. This strategy is analogous to Kripke's account of why someone might believe that it is possible for an a posteriori identity such as 'Water = H_2O' to be false, despite such a posteriori identities being metaphysically necessary truths. In Kripke's cases, we are in an epistemically equivalent situation to being in the presence of water, we are presented with a colourless, odourless liquid which is falling from the sky and filling rivers, lakes and drinking glasses, but the chemical structure of that liquid is different from that of water – it is XYZ (say) – and crucially that substance is not water either (it just looks like it) (Kripke 1980: 97–105).

The details of how a Kripkean explanation is supposed to work in the case of dispositional properties depends upon what else one accepts as part of one's ontology. Specifically, those who accept particulars into their ontology will have a slightly different story to tell from those who do not, and so I will deal with these cases separately. In the literature, it is often presupposed that supporters of an ontology of dispositional properties will reject particulars, and indeed many do so for reasons which should become clear. It is, however, not a requirement of the view and dispositional properties can just as well co-exist alongside particulars (Ellis 2000, 2001).

Those who accept concrete particulars into their ontology in addition to dispositional properties find it easier to adopt a Kripkean explanation: for them, like charges repel each other and do so necessarily because the individuals which instantiate the property of charge must behave in this way. What is occurring in the counterfactual cases, where it *appears* that like charges attract, is that those same individuals (or their counterparts, depending upon your account of modality) are instantiating the property *schmarge* and so the individuals with like schmarge attract each other because of that. So far so good for the Kripkean explanation.

However, this account requires ontological commitment to a species of particulars which some people might find problematic: particulars which maintain their identity independently from the properties which they have. For the Kripkean explanation to work, one and the same particular b must be able to instantiate an actual property P or an alien property Q, but those properties are related to others in their world (as the next response to the problem makes clear). Any account of the identity of particulars which relies upon their having certain properties would fail to ensure transworld identity between particulars in the actual world and those in alien worlds, and so their particularity must be a matter of *haecceity* or this-ness, and a matter of unanalysable fact. In order that this response can offer something different from the one which is to follow, the particulars involved must be characterized independently from whichever properties they happen to instantiate.

Since one of the motivations for maintaining that properties have their causal roles necessarily was to avoid ontological commitment to pure qualities or quiddities, the appeal of haecceistic particulars is likely to be limited among the supporters of dispositional properties. However, it is interesting to note that their acceptance may result in quite a plausible metaphysics, although one which has surprising parallels with the contingentist view of properties to which dispositionalism was intended to be an alternative. Whereas the contingentist about properties locates contingency between a quiddistic property and its causal role (that is, the effect of a property can vary in different possible worlds), the supporter of haecceistic particulars locates the contingency between a particular and the dispositional properties which it has (the properties of a particular can vary across possible worlds). While not literally equivalent to each other, these views provide truth-makers for a roughly equivalent range of possibilities if alien dispositional properties are allowed; but they explain these possibilities by locating the contingency differently in the ontology and both involve commitment to a fundamental category for which only primitive identity criteria can be given.[10]

On the other hand, those who deny the existence of haecceistic particulars, or who prefer to treat particulars as being derived somehow from properties or the entities which ground them, such as being bundles of tropes, have to give a slightly more complicated story about what is happening in worlds where there are alien properties. We cannot say that when we are imagining a pair of particular charged particles behaving in contravention of charge's actual causal role, we are misdescribing those particulars by saying they have charge, when in that possible situation they have *schmarge*. The reason for this is that the particulars themselves are ontologically dependent upon the properties which they instantiate, there is no account we can give of what being the same particular amounts to in isolation from which properties there are, so when the properties change, so do they. In addition, the interactions between those properties will be alien: like schmarges do not attract, they schmattract, and so everything in the alien world is different.[11] Thus what the mistaken person is misdescribing when he conceives of, or falsely believes that, like charges attract is a situation in which *like schmarges schmattract*.

But now the alien dispositional properties case appears to be disanalogous with Kripke's cases of a posteriori necessities. First, the Kripkean account relies upon the mistaken person being in an epistemically identical situation to one in which water is *not* H_2O (say), but no such epistemic duplicate situation is possible in a world of alien properties and laws: one cannot just be in a situation where like schmarges schmattract and the rest of the world is like the actual one, since the causal or nomological relations between properties result in a holistic web of inter-definition. If we are in a schmarge world, then *all* the other properties are different too; the charge world and the schmarge world have no common content and so the claim that one is the epistemic duplicate of the other is, the contingentist claims, far-fetched.[12]

Perhaps the dispositionalist could dodge this criticism: one could argue that holism is not such a serious problem; or that one does not have to accept that dispositional properties entail that holistic system at all. The first option might be achieved by explaining how holistic relations between properties still allow for localized differences to occur between the properties of the actual world and an alien one, differences which are cancelled out by other alterations elsewhere, so that we can be in a schmarge world *w* without this making the whole system of properties and laws differ from the actual one in that world *w*. The second response involves denying that dispositionalism entails the problematic holism: although properties are necessarily connected to each other in lawlike relations, one might deny that all properties are related to all others in this way. If this were true, a local change from actual to alien properties would not result in a global change to an alien world; unlike in the first response, changes in properties could occur without requiring that such alterations be cancelled out elsewhere in the system. But this suggestion might be difficult to accept for many philosophers, especially those of a physicalist or naturalist persuasion: accepting that dispositional properties do not lead to a holistic system of properties and laws would involve denying the unity of science and accepting pluralism about properties, or systems of properties.[13] It would also involve giving up the fundamental conception of sparseness (5.3.1).

I will leave these responses as speculative, however, since another question arises in relation to the dispositionalists' use of Kripke's duplicate epistemic situations. This concerns why we should consider them to be misdescriptions, rather than genuine possibilities as the contingentist argues. Schaffer observes that the strength of the laws which hold between necessarily causal properties is not the same as the metaphysical necessity which ensures that theoretical identities such as 'Water = H_2O' are true, because the lawlike relations between properties concern 'governance rather than identity' (2005: 11). For the accounts of dispositional properties which allow alien properties, Coulomb's Law, or more simply the statement that 'Like charges repel' are false in some worlds; they hold with less than metaphysical necessity, with what we could call nomological necessity. But then why think that there are no worlds in which the actual properties exist although the actual lawlike relations between them fail to hold? The case for this being a misdescription is not strong once alien properties are allowed, since it is not obvious why alien laws which govern our actual properties should be ruled out.

However, Schaffer's objection is not as damaging as it first appears. If properties have their causal roles necessarily, the relations between them will be necessary too. In this, Schaffer agrees, but crucially he presumes that the lawlike relations between properties must hold with a modal strength which is weaker than metaphysical necessity and that this weakens the case for interpreting the conceived contingency of causal role as apparent rather than as a genuine possibility. However, it is not clear why the dispositionalist has to understand the necessity of laws in this way. The dispositionalist can maintain that a specific property has the same causal role in any world in which it is instantiated, while in worlds in which it is not instantiated, the laws in which that property features are still true, but vacuously so; thus, laws hold with metaphysical necessity as they are true in all possible worlds, alien worlds included. This account is in line with Kripke's story: water does not exist in every world, but the theoretical identity that 'Water = H_2O' is still true in worlds where there is no water. If the metaphysical necessity of laws is accepted, then there is as much reason to think that we are misdescribing the apparent possibility of a law being false when the properties within it are instantiated, as there is to think that we are misdescribing situations in which we think that it is possible that water is not H_2O. The disanalogy between the apparent contingency of laws of nature and Kripke's account of the apparent contingency of theoretical identities is illusory.

7.4.3 Counterfactuals

The remaining problem for the dispositionalist is how she will provide truth-makers for counterfactual statements if there are necessary causal relations between properties. If such natural laws are deterministic, the world could not have been different, even slightly different, to the way it actually is without the initial state of the universe being different. For instance, given that I just put a cup of coffee down to the right of my computer, my putting it down to the left could only have occurred had properties been configured differently (and perhaps different properties existed) at the time of the Big Bang; and that seems absurd. So the truth-maker for the counterfactual conditional claim that 'Had I put down the coffee to the left of the computer, the cat would have knocked it over' requires that the universe had an alternate starting point and history. The dispositionalist only seems to be able to account for contingency and thereby provide truth-makers for counterfactuals via alterations in the initial conditions – the properties from which the world is initially 'constructed' – and if we want any change from the actual we have to *back-track* to the beginning of time (assuming, for the sake of argument, that there is such a beginning). The contingentist's possible world, on the other hand, could have been exactly the same as the actual one until shortly before I placed my coffee to the right of my computer, at which point there was a 'miraculous' momentary shift in the laws of nature and I placed the mug to the left.

The dispositionalist has two main responses to this criticism, and one line of attack: she can deny that laws are deterministic; accept back-tracking; and assert that her account of counterfactuals is no more implausible than the contingentist alternative. First, she can deny that the existence of necessary connections between properties entails that the history of the universe follows a fixed deterministic path given a specific set of initial conditions. If causal processes are governed by probabilistic laws – albeit necessary ones – then a change at the beginning of the universe is not required to make the antecedent of a counterfactual true. For each state of the universe (or part of it), there is a range of outcomes which might occur, although one of that range *must* occur. My placing the coffee mug where I did might not have happened because any number of the events which preceded it might not have happened either, its failure to occur would not require the initial conditions of the actual world to be different.

Considered in isolation, denying determinism is a plausible way for the dispositionalist to deal with counterfactuals. But it is metaphysically and methodologically unsatisfactory insofar as it restricts the way the universe could empirically turn out to be, because it rules out a priori the possibility that determinism is true. For those who are uncomfortable about metaphysics placing restrictions upon the range of empirical possibilities, the denial of determinism is not an acceptable response.[14]

The second response accepts back-tracking, and contends that it is no less plausible than Lewis's 'miracles' account of counterfactual truth-makers which allows the laws of nature to be suspended or broken to provide truth-makers for the antecedents of counterfactuals. The dispositionalist accepts that the initial conditions of the universe would have to have been different to how they actually are to provide truth-makers for counterfactuals. For instance, had the initial conditions of the universe been different, I would have put my coffee on the left hand side of the computer and not the right, and so 'If I had put my coffee down to the left of my computer, the cat would have knocked it over' is non-vacuously true. Convoluted though it is to trace contingency and alternate possibilities to different initial conditions, it does not seem less plausible than Lewis's claim that a local 'miracle' (even with the scare quotes) would have been required for me to put my coffee on the left. This favourable comparison with miracles is reinforced given that the back-tracking response is motivated by a desire to sustain the possibility of determinism, and to maintain as wide a range of ways as possible that the empirical world could turn out.

However, the range of possibilities permitted by dispositionalism is still very restricted in comparison to that permitted by the quiddistic account of properties, since properties cannot act in a way which is contrary to their actual causal role; nothing is possible which breaks the actual laws of nature. The properties of the actual world could have been such as to make many facts turn out differently from how they actually are, but they could not have been such as to make the laws of nature different; at least they could not if we maintain actualism and reject a possible start to the universe which included alien properties.

7.5 Categorical and dispositional properties: The debate so far

It seems that dispositionalism can withstand the objections that quiddities are required for an adequate metaphysics. If the dispositionalist does this by accepting particulars into the ontology alongside properties, the result is nearly modally equivalent to the contingentist's theory, except that the former involves haecceistic particulars which may be just as ontologically suspicious as quiddistic properties; while, on the other hand, if the dispositionalist does without such particulars, she must accept that her account of properties places restrictions upon what is possible.

For proponents of dispositional properties, this restriction is plausible and metaphysically worthwhile, since the causal powers of properties can be used to ground both modality and laws, proposals which will be explored further in Chapter 8. While supporters of categorical properties remain unmoved by these supposed theoretical advantages, the current stalemate between the two positions does suggest that there should be no demand to reduce dispositional properties to categorical ones, as long as we are prepared to accept a primitive notion of causality, or natural necessity, in our ontological account of the world. Perhaps if we concede this much however, one might think that categorical properties are the ones which are ontologically superfluous, especially given the sceptical problems associated with quiddistic properties in 7.2.2. Perhaps we should identify properties with their causal powers and reject categorical properties entirely. This is a view known as pandispositionalism and I will briefly explore its plausibility.

7.6 A world of dispositions?

Pandispositionalism raises some interesting philosophical questions and there will not be space to explore these fully here. In what follows, I will talk about dispositional, or necessarily causal properties which lack a categorical aspect as *pure powers*. I will briefly raise objections to an ontology of pure powers which can be avoided by a careful formulation of the theory and then explore two ways in which commitment to dispositions might evolve in order to reframe the relationship between dispositional and categorical properties.

7.6.1 Two regresses

The first objection associated with pandispositionalism is a regress. Pure powers exist when they are dormant, when they are not causally active or manifesting: they have the capacity or capability to act which will only manifest itself in the correct circumstances (to use that term loosely). For instance, a sugar cube has the dormant power to dissolve in water even when it is completely dry. But then a problem arises about how dormant powers ever became active: if all properties are pure powers, then the manifestation of each property must have been triggered by the manifestation of a previous property, but that property must in turn have been activated by the manifestation of a property which was temporally prior to it, and so on; and thus a regress develops. There must be a non-dispositional categorical stopping point, otherwise, there is no way in which the chain of manifestations could have begun and no reason for any pure power to be manifesting rather than being dormant. However, the pandispositionalist might respond that as it stands this regress problem is simply a version of the 'first cause' argument which demands an answer to the question 'In an ontology where all the entities are caused, what caused everything to begin?' Such a difficulty afflicts any ontology in which everything has a cause: we require a first (uncaused) cause to begin the causal process, or else we must accept that the process is perpetual and extends infinitely back in time. The predicament is no worse for the pandispositionalist than for many other metaphysicians, and so this regress does not damage the ontology of pure powers in particular.

However, even if the causal regress is fairly harmless, this objection can also be interpreted as a temporal version of a general conceptual worry about pandispositionalism: dispositions or pure powers alone cannot provide the 'substance' of the world, they are powers, or capacities, but acting, or having potential to act do not (for want of a clearer way of putting it) provide any being. (Or, if powers do provide being successfully, they no longer provide a suitable account of happening (Armstrong 2005: 313–14).) What the regress illustrates, it is claimed, is the need for some entities with a qualitative or categorical nature, at least at some point in time, to provide the stuff or substance of the world, and pandispositionalism cannot provide this. While this problem could be avoided by accepting entities other than powers, such as substrate or particulars, this might be regarded conceding defeat too soon. Perhaps the powers theorist can argue that this objection is driven by unfamiliarity: a world of pure powers seems counterintuitive, but that is simply because it contrasts with our pre-theoretical world view of enduring objects and not because it is conceptually incoherent.

Even if causal activity does provide the being of the world, an additional concern remains about what the existence of dormant powers amounts to: at any moment there is a huge number and variety of powers which are not being manifested. A sugar cube not only has the power to dissolve, but to explode under pressure, to harm a diabetic, to sink a ship, to cause Turkish café owners to break the law[15] and so on. Abundant dormant powers are more uneconomical and mysterious than abundant categorical properties, since their existence is not perpetually 'on show' and will never be revealed to us except if the circumstances are right (and we might expect, given the range of dormant powers there is, that the circumstances will, more often than not, not be right). This objection echoes Armstrong's complaint about dispositions (7.1.1), and the powers theorist can minimize it but not dismiss it entirely by postulating a sparse ontology of fundamental pure powers which determine the rest. The worry about the existence of an uneconomical array of disparate, largely dormant powers is misguided; there are pure powers which are often dormant, but we can accept a minimalist ontology of such entities.

A second regress develops between the trigger of a power – whatever makes that power manifest – and its manifestation (McKitrick 2013). It is often presumed that the trigger for a power P to manifest is the *acquisition* of another power T_1; but then (T_1 having been acquired), in order for T_1 to manifest, the acquisition of another power T_2 is required, but then T_2 (having been acquired) must be triggered by the acquisition of T_3 and so on. This is more serious than the first regress, because the acquisition of infinitely many powers is required within a finite interval of time, between the acquisition of the original trigger T_1 and the manifestation of P brought about by T_1. What is needed to stop this regress is either a power which does not require a trigger, or an account of the triggering of powers which does not involve the acquisition of another power or even the activation of a previously dormant power by means of another. However, McKitrick argues, the former appears to require the existence of non-dispositional properties, while it is not obvious what the requisite mechanism to provide the latter will involve in an ontology where all properties are powers; both options, it seems, go against the spirit of the pandispositionalist ontology.[16]

Broadly speaking, both regresses concern whether an ontology of pure powers needs to be temporally or atemporally grounded by something categorical, substantial or simply not capable of existing in a dormant state in order to be powerful at all. But, first, one might dismiss the suggestion that all powers need to be activated by temporally prior powers – or by something else – in order to be active at all, and accept the powerful ontology as a matter of brute fact. Secondly, the one might give an account of powers which avoids the troubling need for triggers to activate them. Nevertheless, one might still wonder about what being consists in if everything is simply powerful. In the next section I will briefly consider some concessions to this intuition, which might also deal with the problems the regresses raise.

7.7 Categorical and dispositional properties: A distinction dissolved?

Andreas Hüttemann (2013) suggests that the distinction between dispositional properties and categorical properties is not one of kind, but one of degree. As noted above, dispositions are such that they manifest whatever they have the disposition to do whenever the conditions are right. So, for example, the power to dissolve manifests when a sugar cube is put in water, and not while it remains dry. Perhaps there are some powers for which the conditions are *always* right; that is, they are perpetually manifesting. A good example is *shape*: the upper surface of my table is *rectangular* in that it always manifests that particular arrangement of wood. Similarly, the *mass* of the table is perpetually manifesting as well, the table has the mass that it does whatever the conditions. This account of mass and shape brings the paradigmatically categorical properties into the dispositional fold: they are the properties which perpetually wear their causal powers on their sleeves by manifesting them; but they are not fundamentally or ontologically different from those dispositions which only manifest in specific conditions. The only difference between 'categorical' and dispositional properties on this conception is that for the former, constantly manifesting properties, the distinction between a dormant power and its manifestation is philosophically redundant, since the power and its manifestation are always instantiated together.

This account avoids the trigger regress because constantly manifesting properties do not require triggers; they act like categorical properties and thus will always be 'present' or active to provide the triggers which other powers require. It would also provide a constant backdrop of activity which ensured that the world never consisted solely of the potential to act. However, these constantly manifesting properties are not the same as the quiddistic categorical properties which the contingentist favours, and they still count as dispositions on the definition we have employed since they have their causal powers necessarily; *mass* cannot be *mass* and manifest differently from how it actually does. Such properties could be reduced or identified with their causal powers; the ontology is essentially a dispositional one which will inherit whatever restrictions on possibility that involves.

A second conciliatory approach treats properties as having a qualitative *and* a dispositional aspect (Martin 1997; Martin and Heil 1999; Heil 2003, 2005; Strawson 2008). On this view, to conceive of a property as categorical or as dispositional is to partially consider one unitary entity which embodies both aspects essentially and inseparably. If this view is tenable, the arguments which pushed in favour of contingentism arose from an incomplete consideration of the entities involved, and thus our intuitions are faulty. Properties do indeed have their causal roles necessarily, but they also have a qualitative aspect: when we consider this qualitative aspect in isolation, we intuitively suppose that a property can do anything; while if we consider these entities as pure powers, solely in terms of what they do, we might be worried about what makes things what they are. Such intuitions and worries are misplaced. As with Hüttemann's conception of categorical properties as constantly manifesting dispositions, there will be restrictions on the range of possibilities available since this dual aspect view is still a form of dispositionalism: properties still have their causal roles as a matter of necessity and so what a quality can do is fixed.

One might wonder in this case what the qualitative aspect of a property is, since it is doing nothing: Are there different qualitative aspects to go with different causal roles? Or is the qualitative or categorical aspect merely a space-filling extensional nature that is common to all properties which are individuated by their causal roles? (Schroer 2012) The former option involves quiddities in a way in which many supporters of dispositional properties would find uncomfortable (although it does provide a basis for explaining away the contingentist intuitions of 7.3), since the qualitative aspect of a property would be beyond empirical investigation. On the other hand, the latter renders the categorical aspect as uniform across different properties, more akin to a primitive notion of being or substratum from which the causal role cannot be separated. I will not consider here which of these options is the right way of thinking about the dual aspect accounts of properties, except to note that either of them would do something to alleviate problems which pandispositionalism faces. For dispositionalists, accepting that properties have a minimal categorical, qualitative nature may be an attractive concession.

FURTHER READING

For categorical properties:
Lewis 1979, 1986; Armstrong 1999, 2005, Section 3; Schaffer 2005.
For dispositional properties or powers:
Shoemaker 1980; Ellis 2001; Mumford 1998; Marmadoro (ed.) 2010b.

Suggested Questions
1   Are there good epistemic reasons for rejecting contingentism?
2   Can the dispositionalist give a convincing account of what makes it true that 'Had this question been about the distinction between intrinsic and extrinsic properties, it would have been put at the end of the previous chapter'?
3   Do necessary connections between properties create a problem for the Kripkean explanation of why someone can conceive of two masses repelling each other, and yet it not be possible that they do?
4   Are there any non-question-begging metaphysical arguments in favour of either contingentism or dispositionalism? If so, what are they? If not, what would count as a good reason to accept one conception of properties rather than another?
5   Is an ontology of pure powers plausible? Discuss two of the primary obstacles which it faces.
6   Is there a real distinction between categorical and dispositional properties?

Notes

1   I will distinguish necessity and essence to allow that they may not go hand in hand, but I will not take a stance on this (Fine 1994).

2   For an exception, see the strong dispositionalist account of modality (8.5.1) which treats mathematical and logical properties as potentialities, a species of disposition (Vetter 2015: 7.7).

3   I will presuppose that properties can be transworld entities; that is, if one thinks that there are different possible worlds, that an individual property can exist in different worlds. This is a common assumption among those who favour categorical properties.

4   One could attribute the revival to Shoemaker (1980). A detailed history would go outside the scope of this book, but for recent views favouring dispositions see Cartwright (1989), Ellis (2000, 2001), Molnar (2003), Mumford (1998, 2004), Bird (2007) among many others. For an overview see Tahko 2012, and Groff and Greco 2012.

5   We might attempt to analyse statements about dispositions as counterfactual conditionals, for example (Goodman 1954; Quine 1960). The plausibility of this strategy has prompted considerable debate.

6   Recall that this distinction was invoked in order to sustain Lewis and Langton's account of the intrinsic-extrinsic property distinction against the objection that no natural properties would be independent of others (and thus that all properties would be extrinsic) were there necessary connections between properties (6.3.3).

7   Recall that in a material conditional P → Q statement, the whole sentence is true if P is false.

8   If there is no such belief-like attitude, it is not obvious how we could account for the widespread ability to 'believe' in science fiction or fantasy scenarios which, while not prima facie incoherent, might embed inconsistencies that make them so.

9   Such properties may exist either because properties are ontologically-speaking transcendent universals and there is no requirement that they be instantiated, or because possible properties can exist in the same sense as actual ones.

10   More work would need to be done to locate where the range of possibilities permitted by these ontological theories comes apart.

11   It is unclear whether we have to talk in terms of schmattraction in the ontology which includes haecceistic particulars: the particular particles have schmarge which causes them to move towards each other; it is not obvious whether such particulars *moving towards each other* is *schmattraction* rather than plain old *attraction*. This may depend upon the extent to which we have to accept holism about properties which will be discussed presently. However, since haecceistic particulars have their identity independently of the properties they have and can still be identical to those in the actual world, it does not matter to the example whether all the properties are alien in the schmarge worlds or not.

12 Arguably, the Kripkean account of a posteriori identities is in trouble too if the holism in the account of dispositional properties cannot be softened, because a world containing XYZ would not be a world epistemically identical to the actual world were XYZ and H_2O to be dispositional properties, since XYZ is alien and thus related only to other alien properties. Or else, we would be saying that XYZ had exactly the same causal role as water/H_2O, which will lead to the identification of the two if sameness of causal role is sufficient for property identity.

13 Some philosophers would be happy with this world view (Dupré 1993; Cartwright 1999).

14 Some people might think that current physical theory backs up their denial of determinism: the Copenhagen interpretation of quantum theory, for instance, is essentially probabilistic. Nevertheless, one might still feel that it is philosophically unprincipled to accept an ontological theory in which alternative deterministic interpretations of quantum theory could not be true.

15 In September 2014, the governor of Edirne, Turkey banned cafés from serving a second sugar cube with tea.

16 Psillos (2006) presents a related regress against ungrounded powers: that a power P possesses the power Q to manifest, which in turn possesses power R to manifest, and so on. Against which, see Marmadoro 2010a.

8

Causes, laws and modality

This chapter explores the role of properties in the explanation of causation, laws and modality. First, the role of properties in causation is described and the question of whether properties are compatible with singular causation is discussed. Two conceptions of singular causation are distinguished according to whether it is associated with the locality of causation or the particularity of its instances. While several property theories are found to be compatible with the former, only a version of trope theory which rejects exact similarity between tropes is found to be consistent with the latter. Three conceptions of natural laws are discussed – regularity, nomological realism, and dispositionalism – with properties playing the primary role in the dispositional account. The plausibility and sufficiency of dispositionalism about laws is discussed, as are the prospects of its being extended in order to give an actualist account of modality.

The evaluation of a theory of properties will not be thorough without considering the relationship between properties and other entities and processes. The theory of properties is not required simply to classify what kinds of things there are in the world, or to divide them according to what they can do, although these are useful tasks. In most metaphysical theories, properties coexist alongside members of other ontological categories to which they are related in certain ways, such as concrete particulars, or states-of-affairs and are involved in phenomena such as causality, or supervenience, or realization, or nomological connections, which determine what will happen and the structure of the world. Or else, in ontological theories which attempt to restrict which categories there are, such as some trope theories (say), the members of other categories such as concrete particulars or laws appear as derived entities, ontologically dependent upon the more fundamental tropes.

Some aspects of the nature of properties which were discussed in the previous chapters, such as whether the causal role of a property is necessary to it, will have an impact upon the way in which properties can be related to other entities and processes and upon the available conceptions of some other entities, such as causal laws. These implications may be considered as part of the explanatory power of a theory of properties, a prompt to accept or reject it. Furthermore, properties may also have an important role to play in theories of modality, and here too the precise role which properties occupy and their importance within the theory is dependent upon the conception of properties which is adopted. This chapter will examine some ways in which properties fit into metaphysical theories more broadly.

8.1 Causes and effects

A glass window breaks when a brick is thrown at it, a snowman melts when the air temperature rises, Pierre's taking an aspirin cures his headache, and a flock of birds flies away at the appearance of a cat. In our common-sense way of understanding the world – what we might, in this context, call our 'folk-metaphysics' – it is usual practice to pick out or to notice a phenomenon and then to look for what caused it. We often use that cause to explain the phenomenon of interest, or to compare similar cases of causation and to generalize in order to predict future occurrences of similar effects. Where do properties fit into this picture of effects and their causes, which are seemingly connected by a process or a relation which we call 'causality' or 'causation'?[1]

There may be differences between our pre-theoretical understanding of causes and effects and the processes which relate them, based as they are upon our experiences and the way the world appears to us, and what we eventually take the objective reality to be. One might theorize that objectively speaking there is no such entity as the cause, or the effect, since causes and effects are not entities which are objectively distinct from their environment; or that cause and effect might not always be two entities but only one (Molnar 2003; Martin 2008; Mumford 2009); or else that cause and effect are not related to each other by any objectively existing process of the kind we fondly call 'causation'. Nevertheless, properties have a central – often a starring – role to play in many accounts of causality, both the intuitive and the scientifically informed.

An early question which arises when investigating causation is to which category of entities particular causes and effects belong, presuming for the moment that there are such entities. We can consider causation in two ways:

in terms of *general* repeatable relations between entities of similar types (bricks breaking windows, for example); or as *singular* instances of causation (this particular brick breaking this particular window at a specific time, or that brick breaking that window, and so on, each as distinct individual cases, just as tropes are individual qualities). If we concentrate upon singular causation in the first instance, there are broadly-speaking three primary options to be causes and effects: concrete particulars, such as particular objects or events; instances of finer-grained, qualitative entities such as properties or universals, or tropes; or structured, complex entities such as states of affairs or 'facts' where these are considered to be worldly, spatio-temporally instantiated entities, such as a thin particular or substratum instantiating a property at a time.

For example, when a brick being thrown at a window causes it to break, these views give the following ontological accounts of what the cause is:

(i) *Particular Events or Objects*: *the brick's hitting the window* (where this event is spatio-temporally individuated and includes *all* the qualitative features of that region, which may or may not be objectively distinguished into instances of properties).

(ii) *Property Instances*: *the brick's momentum* (say); or its momentum, its hardness relative to the hardness of the window, and an aspect of its shape (say).

(iii) *Structured Complexes*: *Brick b instantiating M at t*, the brick instantiating a property, such as having a specific momentum, at a particular time.

In both (ii) and (iii), properties play an essential role. In (ii) they are sole causes, either alone or in combination with other properties; while in (iii) a property is a constituent of the cause, instantiated by a particular object at a particular time. In (iii), the object cannot be a 'thick' concrete particular – objects like those in (i) which can be spatio-temporally individuated – since such entities include all the properties which the object instantiates; rather, it must be a 'thin' particular, something akin to a substratum which maintains its particularity without itself requiring any qualitative features for its existence (Armstrong 1978a: 114; 1997: 125).[2]

Only in (i), according to which causes and effects are concrete particulars, do properties not overtly play an essential role in causation. But, even on that view, it is commonly maintained that a particular event (or object) has the effect that it does *in virtue of* some of the properties which it has, thereby introducing properties into the causal process as essential constitutive aspects of causes and effects (Kim 1993b). When Pierre takes an aspirin to cure his headache, it is not the shape of the pill, nor its colour that relieves the pain, but its pharmaceutical properties which inhibit COX isoenzymes in cells to prevent the production of the prostaglandins which would have caused inflammation. Although some philosophers do deny that properties are causally efficacious in this way – maintaining that a concrete particular cause produces its effect holistically, in virtue of all its qualitative features (where these need not be and perhaps cannot be, differentiated from each other) – those who hold this view are usually already sceptical about the objective existence of properties (Davidson 1993). The reasons for this scepticism will be explored in Chapter 9; but for now it is sufficient to note that those who deny the central role of properties in causation tend to do so because they have independent qualms about the objective existence of determinate qualitative entities such as properties in the first place. Moreover, in such views, properties (or entities very like them) play an important role – now as part of a theoretical, non-mind-independent ontology – which accounts for causal *explanation* (Davidson 1967).

An alternative paradigm of causation which does not obviously fit into any of the three options above considers the world to already contain, or to be made of 'causal' entities such as *processes*, which are changes or are essentially changing entities. On this ontological picture, what we take to be cases of cause and effect are considered to be processes being moderated, or interrupted, or disrupted from their default course, or else being permitted to resume it (Hüttemann 2013; Dupré 2012). Again, however, something is required to moderate the processes from their default path, a task for which causal powers are postulated; properties thereby make an appearance in the causal mechanisms of the world once again. One might go further and characterize processes themselves as not being fundamental, but being formed or constituted by causal powers or dispositions (Hüttemann 2013). In this case, the ontological account of what causes and effects are would be a version of that in which properties are causes and effects in (ii).

8.2 Singular causation and generality

The central role that properties play in every account of causality which admits objectively existing properties at all is no coincidence, since properties provide the all important link between singular instances of causation and generalizations, regularities or causal laws. As Hume explicitly defined causality, and many philosophers explicitly or implicitly presuppose, it seems that every singular occurrence of cause and effect is an instance of a regularity or law: we can say that *c causes e when occurrences similar to c are followed by those similar to e* (1777, VII.2, 76–7). So, whichever category of entities c and e belong to, there must be grounds for similarity relations between those entities and others of the same ontological category. Properties provide the required ontological grounding for similarity and thus it should be no surprise that they are a common feature of most ontological accounts of causes and effects.

We can characterize the general claim about the relationship between causes and laws as follows:

> *The Cause-Law Thesis*: every occurrence of singular causation is covered by a general law.[3]

If the Cause-Law Thesis is true, then there are broadly-speaking three reasons why that might be:

a) Laws ontologically determine which cases of singular causation occur;
b) Cases of singular causation determine which laws there are;
c) The Cause-Law Thesis is not made true by the ontological relationship between singular causal instances and laws, but is true for another reason.

(a) and (b) make opposing claims about the direction of ontological priority between particular causes and effects and laws, although in both, the existence of one category of entities ensures the existence of the other; while (c) is a catch-all for views which uphold the Cause-Law Thesis but do so either in the absence of ontological dependence between causes and effects and laws (or vice versa), or regardless of such a relationship.

8.3 Properties and singular causation

One may worry that we have simply presupposed the relationship between causality and generality without thoroughly investigating it: first, one might be concerned that we have not clarified the nature of the relationship between singular causation and laws; second, one might argue that common-sense intuitions are faulty and deny that there is an ontological relationship between the two at all.

In order to tackle these problems, it would help to have a clear idea about what singular causation involves, over and above the basic claim that a case of singular causation is an unrepeatable instance of cause and effect.[4] There are two different ways in which we might characterize singular instances of causation as being separate from laws:

The Locality Thesis: the existence of an instance of singular causation is entirely determined by entities intrinsic to it; that is, by the nature of the particular cause and the effect and the relation (if any) between them.

The Particularity Thesis: there are, or could be, unique unrepeatable instances of singular causation which are not instances of causal laws.

The Locality Thesis can be expressed as a claim about the truth-makers of singular causal statements — that the truth-makers of a singular causal statement are intrinsic to the causal relation and its relata — although I have not phrased it that way above, since I do not wish to presuppose that everyone who holds the thesis also accepts a truth-maker principle. The Locality Thesis implies, for instance, that when the like poles of two particular magnets repel each other on a particular occasion, that repulsion is entirely determined by the features of the particular magnets involved and the distance between them. If the Locality Thesis is true, singular causation is a process which is intrinsic to the cause and the effect and the relationship between them, it requires nothing external to these in order to occur.[5] Specifically, it denies the thesis that follows from the Humean definition of causation above which holds that singular instances of causation occur because the properties instantiated by the cause and the effect are involved in regularities; or that they are governed by an external causal law or laws; or (in the truth-maker formulation) it denies that a singular causal statement is true because it is entailed by the existence of a law and facts about the conditions in which the causation takes place. If the Locality Thesis is true, a law may only be essentially involved in bringing about the occurrence of the singular causation if it is intrinsic to that is, wholly present in the instance of cause and effect.

The Particularity Thesis is more radical because it characterizes singular causation in such a way that there might be *no* relationship between singular causal instances and laws: each instance of causation could be a necessarily unique, unrepeatable occurrence, a particular which is ontologically independent of the existence of causal laws.[6] At first glance, the Particularity Thesis might appear to be the negation of the Cause-Law Thesis and thus one might wonder why we need two different theses at all. However, although the Particularity Thesis is consistent with the denial of the Cause-Law Thesis, it is also consistent with its truth in cases of type (c) where the Cause-Law Thesis is true because of something other than there being an ontological dependence between singular causal instances and causal laws. Conversely, the falsehood of the Cause-Law Thesis entails the Particularity Thesis if there is any singular causation at all.

The range of philosophical positions which one can adopt about the truth of these theses is restricted if one takes causal relata to essentially involve properties – that is, if causes and effects are partially or wholly composed of properties – as so many of the accounts of causal relata do. The restriction arises because properties are inherently general, repeatable entities, dividing particulars into types or kinds, and furthermore it is plausible to think that a property's power to cause what it does is shared by all its instances, perhaps necessarily so.

First, let us look at whether the Particularity Thesis is consistent with properties playing an important role in causes and effects. For example, take a singular instance S of causation holding between the *instances* of two properties:[7]

S: p2 caused p5 in circumstances C

Although it may be a matter of contingent fact that S is the only instance of properties *P2 causing P5* which has ever occurred, or will ever occur, such a case does not allow us to say that the Particularity Thesis is true. The reason for this is the essentially repeatable nature of P2 and P5: were another instance of P2 to occur in the relevant background circumstances C, that too would result in the instantiation of P5, because all instances of P2 have the same causal power, in this case to cause the instantiation of P5 when the circumstances are C; S is unrepeated, but not *unrepeatable* as required for the truth of the Particularity Thesis. (Note that it is a trivial matter to uphold the Particularity Thesis if causes and effects are unrepeatable concrete particulars, such as particular objects or events, as long as these cases of causation do not occur in virtue of the properties of the cause; in the latter case, the essential involvement of properties would make such instances behave analogously to the accounts under discussion which have property instances or structured complexes as causes and effects.)

As was discussed in Chapter 7, the question of whether properties have their causal powers necessarily or contingently is disputed. However, to maintain the possibility of genuine singular causation in line with the Particularity Thesis, even contingency of causal role is not sufficient. Although contingentism allows categorical properties to change their causal role, on most views those changes are determined by changes in the laws of nature; so the causal behaviour of properties *cannot* be independent of laws as the Particularity Thesis permits. If, contrary to this position, we were to allow that a property's causal role could change while the laws remained constant, then this would raise questions about whether the same property is being instantiated at different times. Even on the quiddistic conception of properties, it seems implausible that a property could change its actual causal role from instance to instance without corresponding changes in the laws governing them. If properties are treated as ungrounded, or as universals (especially as immanent universals which are wholly present in each of their instances), or as exact resemblance classes, there is no conceptual room to permit a property to change its causal role between actual instances. Instances of such properties gain their qualitative and causal nature in virtue of being instances of repeatable general types and so the individual variation needed to uphold the Particularity Thesis is not permitted.

One might think that adopting a specific account of the nature of properties might help here; specifically, one such as trope theory which permits causes and effects and the causal relation between them to be unrepeatable individuals. However, such individuality will not help if, as was the case in most versions of trope theory, tropes are related by fundamental relations of exact similarity, or they fall into resemblance classes or natural classes, in order to characterize the qualitative similarity between them, since then tropes will be inherently general in an analogous way to other entities which ground properties (see 3.2). This problem will be compounded if trope theorists maintain that membership of such resemblance classes is based upon the sharing of causal role, since then there will be no way for similar cause and effect pairs to behave differently. However, there is one version of trope theory which might be able allow for the truth of the Particularity Thesis, since we could permit an account of tropes in which some tropes are necessarily causally unique. In S*, let p*2 and p*5 be tropes such that:

S*:   p*2 causes p*5 in circumstances C

S* is not much changed from S, except if it is feasible to maintain that the trope p*2 (say) does not, and could not, exactly resemble any other tropes, either with respect to its qualitative nature or to what it can do; that is, that p*2 is necessarily qualitatively or causally unique. This requires that there are *no* relations of *exact* similarity holding between some tropes, such as p*2, although it does permit some degree of resemblance between p*2 and other tropes which falls short of being exact. Thus, p*2 is not of a type such that the existence of tropes like it will cause tropes like p*5 because necessarily there are no tropes like p*2. Only cases such as these – in which instances of properties are genuinely unrepeatable qualities – allow for genuine particular instances of singular causation which are not instances of causal laws. I am not aware of anyone who holds this version of trope theory, but if it stands up to scrutiny, it does permit a version of property theory according to which the Particularity Thesis is true.[8]

In comparison with the Particularity Thesis, the truth of the Locality Thesis is much easier to accommodate in the context of a property-based account of causation. If locality is what is meant by 'singular causation', the property theorist can quite easily present an account of properties which is consistent with it. The Locality Thesis states that singular instances of causation are brought about solely by the entities involved in that singular causal relation – the cause, the effect and the relation between them (which may not be an additional entity over and above the cause and effect) – and thereby it denies that singular causation requires the existence of external entities, such as properties being governed by external generalizations or laws. Theories which uphold the Locality Thesis ameliorate Hume's concern that his definition of causality 'is drawn from circumstances foreign to the cause' since a singular case of causation is only causal in virtue of the behaviour of relevantly similar instances (1777, VII.2, 77).

To give an account of singular causation consistent with the Locality Thesis, a property-based account of causation requires that the entities which are local or intrinsic to the causal relation have sufficient causal power to generate or produce the effect from the cause (whatever that involves). In order to be local, the property instances in singular causal relations would have to be trope-like or to be immanent universals which are wholly present everywhere they are instantiated. Given this restriction, there are two ways in which instances of singular causation can be brought about: in the first, the properties involved are causal in nature, they must be dispositional, or causally powerful, properties which have the intrinsic capacity to do things. Alternatively, the case of properties as immanent universals would also permit general causal laws to be immanent. If causal laws are treated as being relations between universals and the universals F and G are wholly present in the cause and the effect, the law relating F and G – that it is a matter of natural necessity that Fs cause Gs or N(F, G) – might also be wholly present in the singular causal instance (Armstrong 1983). Thus, even if one thinks that properties (the instances of universals F, and G) require a causal law to govern their behaviour and are not inherently powerful themselves, the required law could be local, wholly present in the singular causal relation, and again the Locality Thesis will be satisfied.

One might feel, however, that this latter case of singular causation between immanent universals is consistent with the letter of the Locality Thesis but not true to the spirit of it: the universals involved are not, strictly speaking, local entities themselves, since they are essentially repeatable, and neither is the law which relates them. Thus, this characterization of singular causation is only able to get off the ground by permitting general entities to be wholly present with each instantiation, a notion which some philosophers find conceptually implausible (2.2.2).

It seems that trope theorists fare better in the characterization of singular causation, both in terms of consistency with the Particularity Thesis (with respect to which other ontological accounts of properties do not get off the starting blocks) and the Locality Thesis. Because of the individuality of the entities involved in their ontology, the unproblematic fact of their spatio-temporal location and the possibility that there need not be relations of exact resemblance holding between them, tropes permit a better account of how individual occurrences can be genuinely causal without that being parasitic on the general case. Of course, intuitions about causality are mixed and so the weight which these modest successes will be accorded will vary with them; however, if one cares about having an ontology in which singular causation has a coherent place, or about adopting the ontology of causation which is the least restrictive about such a possibility, this counts as an explanatory gain in trope theory's favour.

8.4 Properties and laws

The discussion in Chapter 7 about whether properties have causal powers, and whether they do so necessarily, was beginning to reveal some inter-esting differences about how we might conceive of properties and the impact which these differences would have upon their relationship with causality and laws. On one hand, properties can be conceived of as intrinsically causally-passive qualities which are acted upon by external laws that govern properties as a matter of natural necessity (Armstrong 1983; Dretske 1977; Tooley 1977), or which fall into general regularities or pattern as a matter of fact (Lewis 1973; Ramsey 1978). Or, on the other hand, causally potent properties can be considered to be the ontological ground of causal laws; the reason why causal relations remain stable over similar cases is that every instance of each specific property involved in them has the same causal power as the property's other instances (Mumford 2004; Bird 2007). On this latter conception, one might even suggest that there are no causal laws as such, except insofar as they are derived from the behaviour of causally powerful properties, and that causal laws could therefore be eliminated from the ontology (Mumford 2004).

   Although this discussion focuses on causal laws, we might also include other forms of lawlike relations between properties which are not tempo-rally ordered, such as *supervenience* or *realisation*. These both involve the ontological dependency of the members of one family of properties upon another, such as mental properties depending upon, or being determined by physical ones, or moral properties being determined by natural or non-moral properties. Like causal laws, supervenience relations between properties support counterfactual and subjunctive claims: for instance, had Pierre not instantiated the physical properties which he does now, he wouldn't be suffering from a headache either. What I have to say here about the different formulations of causal laws will apply to these atemporal relations too, and their formulation and plausibility will consequently be affected by whether the properties to which they relate – especially the 'base' properties which determine the supervenient ones – determine what they do as a matter of necessity.

From the outline above, we have broadly-speaking three accounts of laws:

i) The Regularity Account

ii) Nomological Realism

iii) Dispositionalism

The basic metaphysical differences which motivate and distinguish these accounts can be seen on Table 8.1.

Table 8.1 Laws of Nature

	Is there necessity in nature?	What grounds the necessity?	Which ontological categories are needed for this view?	What are laws?
Regularity Account	No	—	Properties	Regularities or Universal Generalizations: $\forall x(Fx \rightarrow Gx)$
Nomological Realism	Yes	*Connections between properties or their ontological basis (e.g. universals)*	1) Properties (or their ontological basis); 2) Necessary connections relating properties.	Second-order, (nomologically) necessary external relations, between universals (say): $N(F,G)$
Dispositionalism	Yes	Properties/ powers/ dispositions	Dispositional Properties (or their ontological basis).	Internal relations between dispositions/ powers

According to the first account, laws are nothing more than regularities in the distribution of properties, there are no 'real' connections in the world. This analysis of laws has epistemic advantages, since we do not experience necessity (or, at least, it is certainly not obvious that we do), and so it has been popular with empiricists. However, the many regularities which appear to be accidental are a major obstacle to this conception. For instance, consider two chiming clocks next to each other which tell slightly different times but are otherwise accurate: there is a regularity between the first clock striking the hour and the second clock striking the hour, but the former does not cause the latter and we would not want to say that there is a lawlike relationship between the two. One way to rule out accidental generalizations is to demand that laws – which are here effectively statements about the distribution of properties – can support subjunctive and counterfactual claims. That rules out the counterexample of the clocks: if the first clock had not struck, the second would have gone ahead and chimed on the hour anyway; and were the first clock to strike, this does not bring about the chiming of the second. But here we have introduced a requirement to be able to given an account of what is not actually the case, to give truth-makers for counterfactuals and subjunctives. We need an account of the distribution of properties in possible situations in addition to those which are actually instantiated, in order to distinguish between lawlike and accidental regularities.

One way in which we could fulfil this demand would be to adopt Lewisian modal realism (Lewis 1986), but if such a robust account of modality is required to provide an account of laws on the regularity view, we might find it just as convenient to postulate the existence of the requisite necessity elsewhere. This could be in the form of necessary connections between properties, or between the entities which ground properties such as universals, and the latter gives rise to nomological realism (also known as the Dretske-Armstrong-Tooley view, after three of the philosophers who proposed it). On this view, laws are external to the entities which they relate; laws *make* the properties behave in the way that they do. It is consistent with either dispositional or categorical properties, but is usually associated with the latter. (As will become apparent, the dispositionalist does not require governing laws in addition to causally powerful properties to provide lawlike connections in nature and so the combination of the two would be somewhat uneconomical.) However, the nature and the strength of the necessity governing these relations is open to debate: Lewis objects that its nature is opaque at best (1983a: 214), and the best explication which Armstrong provides is to say that the necessity is like that of the causal relation, 'now hypothesized to relate types not tokens' (1993: 422); that is, the necessary connection relates universals (indeed the necessary lawlike connections are themselves universals) rather than their instances. Those who do not regard causality as involving necessity will neither be enlightened nor persuaded by these latter remarks, which leave the nature of the necessity as intuitive at best. It seems that we can only understand the necessity of Armstrong's necessary natural laws by having a prior understanding of natural causal necessity of the variety which the dispositionalist postulates. As for the strength of such necessity, the contingentist about properties will need it to be weaker than metaphysical necessity, in order for it to be possible for properties to change their causal roles, and so nomological necessity is the most likely candidate.

Most important in the context of this book is the dispositional view of laws. If properties are dispositional and have their causal powers necessarily, then there will be lawlike internal relations between them. Properties necessarily determine certain outcomes, or raise the chances of their occurrence (if the world is not deterministic), in a lawlike way, and so nature will appear to be governed by laws of nature while these are nothing ontologically over and above internal relations between the instantiations of dispositional properties or powers. Laws can be treated as ontologically derived entities on this view, and not as a fundamental category in the ontology, and one might think that dispositional properties eliminate the need for them entirely (Mumford 2004).

There are two main questions to consider in order to evaluate the dispositional view of laws: first, whether all laws can be treated as being relations between dispositional properties; and second, what the strength of the necessity is with which dispositional properties bring about their effects and thus what strength of necessity governs natural laws. I will only be able to consider these matters briefly. The first problem concerns whether some laws are unsuited to being characterized in terms of dispositional properties.[9] For instance, one might suggest that some general principles, such as the conservation laws, are too universally applicable to be determined by the causal natures of properties and their instances. These, including the conservation of energy, charge, linear momentum, angular momentum and perhaps others, govern all physical interactions regardless of which dispositions are involved; they are global principles and appear to constrain the way in which properties can interact, and not themselves be determined by dispositional properties. In addition, such principles seem to be contingently true in the actual world; it is possible that the conservation laws are false. If such laws cannot be determined by dispositions, then the dispositional view of laws is insufficient (Chalmers 1999; Lowe 2002b; Fine 2002).[10]

Some dispositionalists have suggested that such general laws are determined by the dispositional properties of whole worlds, such that the world has the property of conforming to the conservation of energy (say). But one might also suggest that *all* actual dispositional properties simply conform to these conservation laws as a matter of brute fact about their causal roles and so certain aspects of causal roles do not serve to draw distinctions of similarity and difference between entities within a world, just as all entities instantiate the indiscriminately necessary properties discussed in Chapter 6. In disanalogy with the indiscriminately necessary properties though, one might allow that there are worlds in which such conservation laws do not hold, and if one favours the response that powers conform to conservation laws in virtue of their intrinsic natures, there will have to be different, alien powers in such worlds. This consequence seems of a piece with the holism associated with dispositional properties though: it is possible that the conservation of linear momentum fails, but then it is plausible to say that *mass* does not exist in such worlds, there is another alien property instead. The dispositionalist need not hold that it is contingent of the properties in the actual world that they conform to the conservation laws.

The second issue concerns the strength of the necessity with which dispositional properties bring about their effects, which does not merely have consequences for what laws of nature are like, but also for whether the ontology of dispositional properties can also be employed to explicate modality. Opinion is divided between those who hold that dispositional properties have the effects they do as a matter of metaphysical necessity, or nomological, natural or conditional necessity, and those who maintain that properties act with dispositional necessity in having only a tendency to bring about their effects, even when the circumstances are right.

There is much to be said about whether it is coherent to think of dispositional laws as holding with necessity of a weaker strength than metaphysical necessity. The principal obstacles are associated with ensuring that conditional, nomological or dispositional necessity do not collapse into contingency on the one hand, or metaphysical necessity on the other, but I cannot do justice to that debate here.[11] Nevertheless, we have encountered some of these alternatives and the arguments for them before. During the discussion of whether properties are dispositional in Chapter 7, it became clear that the dispositionalist would be better served if the relationship between a property and its causal role is metaphysically necessary: only if she considers necessity univocally that is, metaphysical and nomological necessity collapse into necessity of a single strength do the intuitive metaphysical arguments in favour of dispositional properties have any traction (7.2.1); and, furthermore, the Kripkean explanations about how we might mistakenly think that certain scenarios were possible when they are not would also be strengthened were the laws of nature to hold with metaphysical necessity (7.4.2). There are drawbacks with this view, some of which appeared in 7.4.3 and others which will be considered in the next section, with the primary complaint being that if dispositional properties bring about their effects with metaphysical necessity, then it is difficult to make sense of counterfactuals, with the intuitive notion that things could have been different from how they actually are. In addition, one might worry that the non-actualist versions of this view which permit the existence of alien dispositional properties, such as *schmass* and *schmarge*, and therefore alien laws such as the alien inverse cube law which governs the schmattraction of schmasses accept an account of metaphysically necessary laws whereby many alien laws are vacuously true in the actual world. The inverse cube law prevails in the actual world if laws are metaphysically necessary, but it is never instantiated because there is no schmass. This is, for some philosophers, a counterexample to this view of laws (Fine 2002: 260), but it is not clear why the proliferation of laws should be any more troubling for the dispositionalist than the vacuous truth of myriad uninstantiated theoretical identities, such as 'Twater = XYZ', is on the Kripkean view which regards them as a posteriori metaphysical necessities. Once one has embraced alien properties and given up actualism, the consequences of permitting any metaphysical necessities leads to an abundant but mainly redundant stock of properties and laws.

One might prefer laws to be nomologically necessary on the basis that there are possible worlds in which the laws do not apply. But although this deals with the proliferation of vacuously true laws permitted by the previous account of laws as metaphysically necessary, it seems to be a trivial, almost terminological concession which is as yet not well motivated: we have only a primitive conception of what nomological necessity is and the unified account of necessity fares better against the contingentist's counterarguments of Chapter 7. Furthermore, for dispositionalists, the more significant metaphysical decision is not between nomological and metaphysical necessity, but about whether they are actualists or not; that is, whether they countenance the existence of alien properties and thus alien laws, since a property will determine its causal outcome in a lawlike way in every world in which it is instantiated. We can call that either 'nomological' or 'metaphysical' necessity as we wish, but it will not change the strength of the causation and thus the law. The important decision is whether properties are restricted to those which exist in the actual world.

8.5 Properties and modality

Just as the different conceptions of properties can influence how laws of nature are characterized, so too can they alter the account of modality which can be given. For those who want to provide an account of the truth-makers of modal statements – statements about what is possible, or necessary, about what could happen, could have happened, or must happen, and so on – there are several ontological accounts of what determines possibility and necessity. There will not be the opportunity to discuss these different views here, except insofar as they are influenced by specific accounts of properties, and thus their success or failure may count in favour of, or against, a version of property theory. So the accounts below should not be treated as an exhaustive survey of the metaphysical explanations of modality on offer.[12]

First, it is worth noting that any attempt to explicate modality in terms of properties runs the risk of being caught in a tight circle of interdefinition since, in several cases, the characterization of properties has presupposed modal notions. For instance, if properties are constitutively identified by being necessarily coextensive – by their being instantiated by the same set of possible and actual particulars, say – then the range of possibilities must be ontologically prior to the existence of properties, or at least ontologically interdependent with it. Similarly, if one thinks that properties are constitutively identified and individuated by their causal roles, or their role in laws, then properties cannot be used to elucidate what is nomologically possible or necessary either. These accounts would not beg the question however, since we are concerned with constitutive criteria of identity, rather than epistemic ones; we are not interested in how we might determine whether properties are numerically identical or different, but in what makes them so. So a circular theory points to the interdependence of ontological categories, rather than a vicious circle of the epistemic kind. Nevertheless, an account which explicates modality in terms of entities which are themselves determined by the range of possibilities there is will be short on explanatory power. The property theorist faces a dilemma here between being able to provide constitutive criteria of what makes members of the category of properties the same or different, or being able to use those entities to provide an account of modality.

From the discussions in 5.2, it is not obvious how properties can be individuated in a sufficiently finely-grained way without recourse to either metaphysical or nomological modality. So, an independent account of modality which does not presuppose the existence of determinate properties will be needed by those who make identity criteria a requirement for the acceptability of an ontological category. The alternative is to regard which properties there are as primitive, which leaves the way open to explicate modality in terms of them.

As with the metaphysical accounts of laws of nature discussed in 8.4, the most strikingly different accounts of modality arise as a result of the distinction between categorical and dispositional properties. Categorical properties can be used to ground a combinatorial account of possibility with each (maximal) permutation of properties constituting or representing a possible situation, or a way the world could qualitatively be. However, since categorical properties have their causal roles contingently, what can happen in these possible situations will not be determined by the ontology of properties alone, but will require properties to be governed by laws of nature in each world. Alternatively, if causal laws are determined by regularities between properties, then the distribution of properties in a world and others qualitatively similar to it will determine what counts as an instance of causation in that world.[13] Although properties play an important role in combinatorialism, and a lesser one in modal realism, these accounts of modality both require the existence of other fundamental categories – laws or possible worlds respectively – in order to determine what can happen; properties are not able to ground modality on their own.

Dispositional properties, on the other hand, provide a ready-made ontology of modal entities which determines how things could possibly be (Borghini and Williams 2008; Vetter 2015). What dispositional properties can do is part of – perhaps the whole of – their intrinsic nature, and so the existence of certain properties necessitates the existence of others. This determines what can happen and what types of things can exist by placing restrictions upon the combinations of properties which can co-exist, narrowing the range of possibilities in comparison to those available with an ontology of categorical properties in which any necessary connections are externally imposed. It is important to note that the dispositional account does not attempt to reduce possibility to something else, since properties have an irreducibly modal nature, the causal role which determines what they can do.

Cross-cutting this distinction between the combinatorial and dispositional ways of understanding modality is another distinction, based upon whether the account is *actualist* or not; that is, whether what is possible is entirely determined by what exists in the actual world, in this case, which actual properties there are. In property theory, we can also draw this distinction by noting that the actualists adhere to the Principle of Instantiation which restricts the domain of properties to those which are, have been, or will be instantiated in the actual spatio-temporal world, and rules out alien, uninstantiated properties. Note that this restriction does not rule out an account of properties as abstract, transcendent universals which nevertheless conform to the Principle of Instantiation such that only instantiated universals exist.

Rejecting actualism and the Principle of Instantiation involves accepting the existence of some purely possible entities – in this case, some alien properties – and thus introduces a prior commitment to modality. The only available answer when we ask which alien properties there are seems to be 'all the possible ones'; the non-actualist needs an independent explanation of which possible properties exist, such as those which exist in Lewis's possible worlds, or he must treat the matter of which alien (and actual) properties there are as primitive. While there is nothing philosophically wrong with such a primitive assumption and in fact, a similar assumption is made about the domain of Lewis's possible worlds, the property theorist cannot then claim to have analysed modality in terms of properties. For this reason, I will concentrate upon accounts of modality which take the actualist route.

8.5.1 Modal dispositionalism

There are some significant differences between the accounts of modality available to the dispositionalist and the combinatorial actualism available to the proponent of categorical properties. While the dispositionalist is able to give an ontologically parsimonious account of possibility, by presupposing the existence of only one category of entities and actual ones at that, her theory does not permit as wide a range of possibilities as the contingentist combinatorial alternative. In particular, as was noted in 7.3, the dispositionalist is unable to account for possible worlds which involve the same properties as the actual one but in which the laws of nature are different. (Nor, if she is committed to actualism, can she accept a world in which the properties are different to those in the actual world.) On the actualist dispositionalist view, nothing is metaphysically possible which is not also nomologically possible – the distinction between metaphysical and nomological possibility collapses – since everything which could happen is determined by the intrinsic causal natures of properties.

One might object that this account of modality is too restrictive, and there are two responses which the actualist dispositionalist can make. The first is to embrace the merging of metaphysical with nomological necessity and the restrictions upon possibilities which that involves (Shoemaker 1980; Swoyer 1982; Ellis 2001; Bird 2007; Wilson 2013). As we saw in 8.4, it is plausible to accept the metaphysical necessity of the nomological connections between dispositional properties for the perspicuous account of laws it offers, and the fact that possibility is univocally explained in terms of actual, non-abstract entities. However, this ontology restricts the range of counterfactual situations for which we have truth-makers and reduces our ability to explain them away. For instance, while we might plausibly give an account of how counterfactuals might have been true by either denying determinism or backtracking to accept different initial conditions (7.4.3), this strategy will not work with counterlegal claims, such as 'If the law of gravitation did not hold, my pencil would remain in the air when I let go of it'. There is no way in which we can reorganize the initial conditions of the world, so that a law of nature turns out to be false, nor can the introduction of probabilistic outcomes help the case. What should the dispositionalist say now?

The dispositionalist might try to turn the objection to her advantage by emphasizing how problematic our epistemic position is with respect to the objector's purported counterlegal possibilities (Bird 2002; Wilson 2013). Can we really assess a counterlegal claim such as 'if the law of gravitation did not hold, my pencil would remain in the air when I let go of it'? In such a world (which is nomologically impossible but metaphysically possible for the contingentist and impossible for the dispositionalist), it is difficult to say what would happen and which entities there would be; it is unlikely that there would be any pencils to drop, for example, nor anyone to drop them. We simply do not have an epistemically reliable grasp on what a world containing the same sorts of things as the actual one which had different laws would be like and how the entities within it would behave. The actualist dispositionalist account of modality is sufficient, it is argued, to provide truth-makers for a wide range of counterfactuals, and we should not worry that counterlegal statements turn out to be trivially true because their antecedents are necessarily false.

The second, alternative response is to provide a greater range of possibilities while maintaining actualism by postulating the existence of a greater number of actual dispositional properties than we would ordinarily think exist, or by reconceiving the way in which dispositions are possessed. For example, Vetter (2011, 2015) suggests that we can think of objects as being disposed to do things to differing degrees: for example, a spun-sugar sculpture is (perhaps) *maximally* fragile, a champagne glass slightly less so, a piece of perspex less so still ... until at the other end of the scale we might talk of an (almost) minimally fragile tonne of lead, or gold, or plasticine. Maximal possession of a disposition rules out the opposite potentiality being possessed. Vetter calls such dispositions which are had by degrees *potentialities* and they are, she argues, sufficient to provide the grounding for modal claims which are not catered for by the ordinary dispositional properties which are manifested around us.

This second response involves populating the actual world with sufficient truth-makers to account for many of the counterfactuals which the actual world appears to preclude. While simple backtracking can provide us with unicorns or talking donkeys, if we go sufficiently far back in time and let the properties determine a different causal path, the manifestation of potentialities which are only possessed to minimal degrees can cater for more outlandish possibilities. Plasticine might have shattered when I hit it with a hammer (or else the hammer shattered when I hit it with the plasticine), unlikely though these potentialities are to ever manifest. However, despite the increased modal range permitted by this more populous ontology, it is less clear that Vetter can accommodate the possibility of the actual laws of nature being false, especially when these concern relations between fundamental potentialities (dispositions) which we cannot backtrack to 'get rid of'. One solution to this problem might be to permit the *world* to have potentialities, in order that it could have manifested in a different way to how it actually is. But, since potentialities are supposed to be a species of dispositional property, it is not clear what kind of entity the world must be to have potentialities of its own (Vetter 2015: 7.3). At best, this solution is underdeveloped and requires a conception of the world such that it can have potentialities which are not possessed by, nor derived from, the actual potentialities which make it up.[14]

Vetter's solution to the problem of counterlegal statements relies upon her conception of potentialities being had by degrees, and the crux of the matter depends upon whether laws of nature turn out to be necessary or not on her view. When like negative charges repel according to Coulomb's Law, it is not clear that the possession of electric charge should be identified with having maximal potentiality to act in a certain way, or simply with having that potentiality (to some degree or other). If the former, Vetter argues, then the laws of nature do indeed turn out to be necessary, but we are owed an explanation of why it is that everything which has charge only ever has it maximally; without such an explanation, the principal reason why the laws are necessary has not been explained. On the other hand, in the latter case, the way is open for Coulomb's Law to be violated: an object which has the *potentiality to do P* less-than-maximally also has the opposite potentiality not to behave according to P, and thus not to instantiate the laws or regularities in which P figures.[15]

The plausibility of Vetter's response depends upon the specific conception of dispositions or potentialities in play – namely that their possession may be a matter of degree – and whether this permits our taking laws as holding with less than metaphysical necessity. Secondly, it requires that we accept that the range of dispositional properties which any particular possesses far outruns those which we might ordinarily think it has, even under enthusiastic scientific investigation. The plasticine is fragile, and the donkey has the potential to speak French, although neither have these potentialities to a very great degree. The plenitude of potentialities on this view might be thought to pose a problem and to remove some of the attractive parsimony of the dispositionalist account of modality.

Furthermore, this problem appears to be magnified if the dispositionalist chooses to formulate what has become known as 'strong dispositionalism' and to attempt to accommodate logical and mathematical necessities dispositionally too. This would provide a more complete account of modality using only dispositions, but it prima facie involves every actual entity (maximally) having uncountably many dispositions which correlate with the indiscriminately necessary properties discussed in Chapter 6. For instance, each and every thing has the potentiality *to dance or not to dance, to be such that a square exists, to be such that 2+2=4,* and so on. The plenitude problem is getting worse.

One might accept at this point that it would be preferable to accept a dualistic account of modality: dispositions or potentialities are fine for dealing with the causal possibilities of the actual world, but we can accept that logical and mathematical necessities are different in kind. (We can call this view 'weak dispositionalism'.) On the other hand, one might protest that the necessarily manifested logical potentialities are harmless enough – I am, after all, either dancing or not dancing at any point in time, as is the desk upon which I'm writing, so there is clearly a potentiality to do so in both cases – while mathematical potentialities can be treated as being necessary manifestations of a much narrower range of entities, namely whatever the truth-makers of mathematical statements are (which may vary according to your philosophy of mathematics). Thus, one could sustain the view that actual potentialities are all that are required for a complete account of metaphysical modality; possible worlds, either real or considered as abstract objects, are no longer needed. This view might be regarded as an advantage of the dispositional conception of properties which made its formulation possible.

Yet one might argue that Vetter's strong dispositionalism is successful precisely because the conception of a disposition which she uses is different from our usual understanding of dispositions or dispositional properties. The super-abundance of actual potentialities which each object possesses only to a minimal degree serve to widen the modal range which the actualist dispositionalist ontology can account for, in order to provide for all metaphysical possibilities without reducing the metaphysical possibility to nomological possibility narrowly understood. But one might be concerned that such potentialities are just as strange as the purely qualitative quiddistic properties which can change their causal role, or the alien properties which the dispositionalist could postulate were she prepared to abandon actualism in pursuit of a greater modal range. The fact that the strong dispositionalist ontology is entirely instantiated in the actual world, and does not require the additional ontological baggage of possible worlds, might still stand in its favour. But nevertheless, we are being asked to accept that each object intrinsically possesses an abundance of potentialities which are never manifested, many of which are directly opposed to the properties of the object we observe.

8.6 Conclusion

This chapter has been something of a whistle-stop tour through some of the primary uses to which an ontology of properties might be put. Many controversies remain undecided and objections unconsidered.

Nevertheless, from the point of view of providing a perspicuous account of causality and laws, the dispositionalist option is a clear winner when judged on ontological economy and it also permits one to provide an account of modality as well. However, there are three points which should be noted which may count against it: First, it involves significant ontological commitment to entities which have instrinsic natural necessity. As was noted in 7.1.1, this is an epistemologically uncomfortable position since no amount of empirical investigation will reveal the existence of such necessity; rather, we must presuppose it for the explanatory work it can do. Second, even on the most abundant account of actual dispositions, Vetter's potentialities account, there may still be limitations upon the range of possibilities which dispositionalism can account for. Third, although dispositionalism may appear to be doing better than theories of categorical properties, and the additional ontological categories required, to give an account of causality, modality and laws, these phenomena form a fairly small part of what we want to metaphysically explain. The additional ontology needed by the contingentist – which is here being branded 'uneconomical' – may turn out to have explanatory utility in other areas of philosophy; we should not reject the less economical theory out of hand.

FURTHER READING

On laws of nature: Armstrong 1983; Beebee 2000; Mumford 2004; Bird 2007; Schrenk 2010.
On modality: Borghini and Williams 2008; Vetter 2011, 2015; Mumford and Anjum 2011.

Suggested Questions
1 What is singular causation? Is any property theory consistent with it?
2 The acceleration of a mass in a frictionless environment is proportional to the force applied to it. What makes that generalization true?
3 Why are properties important in the explanation of causation?
4 Can an actualist dispositional account of modality give a plausible explanation of what makes the following statements true? Does it matter to her theory?
    (i) If the Earth's atmosphere had been like that on Venus, intelligent life would not have evolved.
    (ii) If the change in gravitational attraction in relation to distance was described by an inverse cube law, rather than an inverse square law, the planetary orbits would not be stable in 3 dimensional space.
    (iii) It must be true that 3+3=6.
5 Does a dispositionalist require laws of nature as well as properties?

Notes

1   I will use 'causality' and 'causation' interchangeably. There are numerous philosophical questions concerning causation and I will barely scratch the surface. See Sosa and Tooley 1993.

2   Armstrong seems to regard the distinction between thin and thick particulars as conceptual, rather than ontological: the thin particular is what we get when we consider a particular in abstraction from all the properties which it instantiates; that is, when we partially consider the particular, focusing on its particularity (1997: 123). One might take a more ontologically robust view of the distinction, however.

3   The name is from Davidson (1995), who holds a version of (c).

4   A thorough investigation of singular causation would go outside the scope of this book, but see Ducasse 1926; Anscombe 1971; Davidson 1967, 1995.

5   One could allow that what counts as intrinsic to the singular causal instance includes the particular conditions within which the causation occurs, in order for this thesis to be compatible with the view that causes and effects cannot be determinately isolated from their particular environment.

6   As above, one can allow that the individuation of singular causal relations might involve abstraction: causes and effects might not be discrete entities which are objectively determinate or separable from the background within which they occur.

7   I will use this simple case, but this example could be extended to apply to causal relata which are composed of properties such as states of affairs, or Kim's property-exemplifying events (Kim 1992; Armstrong 1997).

8   Cartwright's account of capacities having causal power, such that there are real instances of causation between capacities, while causal laws are instrumental and not objective, is consistent with this view although Cartwright has not (to my knowledge) presented the metaphysics in detail (1980, 1983, 1989).

9   This is not the same problem as that of some plausible laws not being determined by actual properties; we can, if such examples sway us, allow for alien properties which will determine alien laws.

10   The regularity view also has difficulty giving an account of global patterns of behaviour among properties; it is not clear why every local collection of properties apparently conforms to such general principles if they are not governed by a general overarching law.

11   I will skip over this issue with regret because the intricacies of the debate would require at least another chapter. For further discussion see Marmadoro 2010b; Mumford and Anjum 2010, 2011; Schrenk 2010; Wilson 2013.

12   For an excellent survey of the subject, see Divers 2002.

13   As noted in 8.4, the regularity account of laws makes most sense on a modal realist view to deal with accidental regularities, but this already

involves a substantial account of modality in which properties play only a minor role.

14 Attempts to accommodate global principles such as conservation laws by arguing that the world has dispositions face similar difficulties (8.4).

15 Vetter attributes a similar view of laws to Cartwright (1989), Schrenk (2010). As noted in 8.4, there is much to be said about the modal strength of dispositional laws which will not be covered here.

9

The ontological status
of properties

Are there mind-independent properties? If there are, do we know which ones exist? First I investigate whether these questions make sense by exploring some methodological attitudes to metaphysics. I argue that there are good reasons to formulate and assess theories of properties, even though one may not believe that there is a fact of the matter about which one is correct. I distinguish two theses about the existence of sparse and abundant properties, and argue that acceptance of the former is motivated by its utility in metaphysical explanation, and there may be reason to hold the latter on scientific grounds. Having accepted the existence of mind-independent sparse properties, I evaluate sceptical arguments which raise difficulties about our epistemic access to properties whether such properties are quiddistic, or dispositional. Finally, I sketch what theories of properties might look like if properties are not entirely mind-independent entities.

In previous chapters, I have presupposed that at least some properties are mind-independent entities, qualities which exist objectively, whether or not any conscious beings exist to take any notice of them, to try to discover them, to individuate them, or to use them to classify the things they encounter. While we have been concerned with metaphysical questions, this ontological attitude has been harmless enough, since the aim has been to seek out the best formulation, or formulations, of a theory of properties, and questions concerning their ontological status have not seemed relevant. This chapter will explore this presupposition about the objective existence of properties in more detail, question the reasons for holding it, and explore some of the alternative views about properties which one might adopt if it is found wanting.

Before starting this discussion, we should note that the ontological status of an entity or a category of entities is not something that it can have or lack; that is, ontological status is not to be identified with an entity's reality, or its objective existence, although the term is sometimes used in this way. The 'ontological status' of something, in the way I will use it, can roughly be cashed out as 'how it exists' or 'what status its being has'. Such a gloss is precarious and I do not intend to mean too much by it, except to emphasize that on this dangerously broad interpretation of 'existence' or 'being', everything has existence of some kind or another; an entity might be real, or instrumental, or linguistic, or mind-dependent, or ideal, or imaginary, and so on, with these terms meant to describe its ontological status. I will not attempt to give an in-depth account of the options available, although I will explain them as they arise. In addition to this, I will mainly use 'exist' and related terms in their more restricted usage which only covers a subset of existence in the broader sense just given; that is, I will avoid talking about the 'existence' of imaginary entities, or ideal entities, and the like, without additional qualification.

9.1 The existence of properties: Property realism

Does the world contain properties independently of us? Or, to put the point another way, does the world have objective qualitative joints? We might think that it is intuitively obvious that it does: our everyday language and our scientific theories abound with attributions of properties, the confirmation of laws or regularities which seem to link them, and counterfactual judgements which say how things would have been had the actual properties been different. We cannot avoid classification and generalization, and judgements of qualitative similarity and difference underpin other judgements concerning persistence and change, they underlie our actions and the difference they might make. It seems natural to think that such judgements – at least some of the time – are about the way in which the world is put together. It is no coincidence that accounting for objective qualitative similarity and difference between distinct particulars is one of the common-sense facts which Moore thought it essential that philosophy explain (1959).

But, these collected observations do not amount to a deductive argument – or even the bare bones of an argument – in favour of the objective existence of properties, since we might be able to continue all the activities listed in the above paragraph and yet our qualitative judgements be dependent, at least in part, upon us and our way of thinking. We have a sceptical problem: How can we know whether the properties we observe exist independently of us? But this scepticism does not just apply to properties, but to everything which exists mind-independently; it is a general problem for realists. Perhaps scepticism about properties is no more serious than common-or-garden radical scepticism and can be dismissed or ignored accordingly. Both the realist about properties and the sceptic have some work to do to make their respective cases.

9.1.1 Two questions

We can begin by asking two related questions: Do properties exist? (That is, to be clear: do they exist mind-independently?) And, if properties do exist mind-independently, how do we know which properties there are? If the first question is answered in the negative, then the second becomes obsolete. But even if there is good reason to give the first question an affirmative answer, the answer to the second question may simply turn out to be that 'we don't know which properties there are', if a plausible kind of scepticism holds sway. Even when we soften the epistemic demand to find out about properties to our having justified belief about which ones there are, it may turn out that our epistemic access to mind-independent qualities is insufficiently robust. I will discuss the first question and later in the chapter move on to the second.

9.1.2 Do properties exist? Two principles

Given the different ontological accounts of properties we have explored, there are correspondingly different ways to characterize the claim that these entities exist objectively or mind-independently. However, it is possible (and prudent) to simplify these into two distinct theses, the first maintaining the existence of an elite set of fundamental sparse properties, and the second broader claim which maintains the existence of a more populous ontology than the first. We can formulate these as follows:

The Sparse Properties Principle (SPP):[1] there is a set of sparse properties existing objectively or mind-independently.

The Abundant Properties Principle (APP): there is an abundant collection of properties existing objectively or mind-independently.

In the current context, it is important to note that both principles are blind to whether properties are spatio-temporal or abstract entities, as long as they exist mind-independently, and the formulations are also neutral about which ontological basis of properties (if any) is preferred. Secondly, the principles are not mutually exclusive: one could be a dualist about the different species of properties and maintain SPP and APP concurrently; or, given the existence of a suitable generating mechanism for combining sparse properties, one might maintain that sparse properties determine abundant ones and so SPP entails APP (4.7, 5.3). In the latter case, APP would not be a basic fact about the ontology and would not require an independent argument for it. Furthermore, SPP would not result in a less populous ontology than APP, just a less populous fundamental ontology.

9.1.3 Mind-independence

The term 'mind-independent' is not intended to exclude properties which are *instantiated* by particular minds, such as beliefs, desires, or properties of consciousness. Although such properties could not exist were minds not to exist, one might still think that the distinctions between these properties mark objectively existing joints in nature (although they are unlikely to be sparse properties). 'Mind-independent' in the sense I intend should be taken to include psychological properties too, if a mind or minds do not play a role in identifying, individuating or postulating such properties.[2]

Furthermore, although there are extremes of entirely mind-independent entities on the one hand and entirely mind-dependent entities on the other, there is significant scope for middle ground between these two extremes. It seems plausible to think that an entity which is not entirely mind-independent is not thereby wholly dependent upon the existence of a mind, since it may be only partially mind-dependent. So, although the denial of realism which I am characterizing in terms of mind-independence is often equated with idealism, or anti-realism, I do not intend to do this. Unless an argument presents itself to rule out my position, I will suppose that an entity, or a category of entities, can be dependent upon both the objective world and the mind, thereby occupying a graduated middle ground between the two extremes.

9.2 Two radical approaches to ontology

9.2.1 *Quining properties*[3]

One might already think that mind-independent properties are more trouble than they are worth. Do we really have a firm grasp on what qualities are? As discussed in 5.2, the provision of identity and individuation criteria for properties quickly becomes dependent upon the existence of entities more obscure than properties, such as possible individuals or worlds,[4] until the project appears to break down entirely. It is difficult even to assess the different proposals fairly, since we do not seem to have a clear idea of how many properties there are – that is, how finely grained the divisions between qualities are – without reference to the theoretical purpose properties have, but there are choices about what role properties play in the ontology. We seem to be caught in the firm grip of intuition when we try to answer questions about what properties are.

If one takes the minimalist approach recommended by Quine, properties – or 'attributes', as he calls them – do not pass the most basic of tests for admission into the ontology. Nor do they pass Quine's second test of ontological respectability: they are not entities which we quantify over in our best scientific theories (1948). One might contest this point: Do we not quantify over properties in science? I think the answer to this is often negative. While one might presuppose that the predicates of our scientific theories pick out properties in the natural world, we can formulate our scientific theories to do without specific reference to the category of properties itself and frequently do so. *Properties* as a metaphysical category do not have an indispensable position in our science, so there is no reason to think that they must exist.[5] On Quine's view of ontological commitment (although it is open to interpretation),[6] entities which exist according to one theory may not be the entities which exist according to another; ontology is relative to the theory we are in. So the appearance of properties in some theories – even our best, most complete theory – does not justify our taking a realist view of them. As far as Quine is concerned, we should accept that there is nothing more to be said about why particulars have the properties that they do, or why predicates apply to them. (This approach is known by the pejorative 'ostrich nominalism', from Armstrong 1978a.) Quine sees no reason for ontological commitment to objectively existing determinate properties.

9.2.2 *Carnap's questions*

While Quine's approach counsels against commitment to mind-independent properties, Rudolph Carnap's attitude to metaphysical theorizing recommends a different kind of rejection (1947). Rather than our not having reason to commit to the entities postulated by SPP and APP, Carnap takes issue with asking about the truth of the principles themselves. He distinguishes two ways in which we can understand questions about existence, such as 'Do properties exist?'. On one hand, he argues, properties clearly do exist, just as numbers obviously exist when we do mathematics: we talk about properties; we attribute them to objects; we use them to classify things; and this is a book about them. There is a sense in which it is trivially true that properties exist and it makes no sense to deny that there are properties. We make this judgement that properties exist from within our common-sense and scientific view of the world – the existence of properties is trivially true from *within* our theory or worldview – and so Carnap calls questions about existence which have trivial answers because of the theory that they are asked within *internal* questions. On the other hand, if we try to 'step away' from the theory and ask a question about whether a category of entities *really* exists, independently of any theory which includes them, then the question is unanswerable because the answers have no truth value. Questions concerning what exists independently of a theory he calls *external* questions, and these are questions which it makes no sense to ask.[7]

If Carnap is right, there is no point in asking about the truth of SPP and APP, because they are best interpreted as being claims about the existence of a category of entities independently of our minds and any theory we might have; as such, they are meaningless, or at least unanswerable questions. On the other hand, asking about the existence of properties in general, without seeking to ask about what is true independently of a theory, leads to the obvious and trivial answer that properties do exist. We can accept the existence of properties if they appear in the theory that we're using and reject them otherwise; there is no point in worrying about whether properties 'really' exist mind-independently because there is no fact of the matter about whether they do or not.

The Carnapian approach to metaphysics, which has been recently revived, would seem to dismiss the concerns raised in this section as a whole (Hirsch 2002, 2009; Chalmers 2009). However, that conclusion is too hasty, since there are some theories which include supposedly mind-independent, fine-grained properties as part of their ontology, and others (such as Quine's), which do not. Those who sympathize with Carnap will still have to make a principled decision between competing theories and so, even if we accept the neo-Carnapian view of ontological questions, we still need a way to decide between one theory and another.

The differences between the ontological realist and the Carnapian anti-realist will begin to show again should we find theories which are as good as each other; that is, they can explain all the same phenomena, or almost all the same phenomena (and have comparable explanatory strengths in other areas). In such cases, the ontological realist will insist that there is a fact of the matter about which theory is the true one, although we may never be able to know which one that is (Bennett 2009), while the Carnapian view will either be that there is no fact of the matter about which theory is right, or that the theories are terminological variants of each other. Given their internal consistency and equivalence in explanatory tasks, there is a good argument to suggest that some of the different ontological accounts of properties discussed in Chapters 2, 3 and 4 are explanatorily equivalent in this way.

While a Quinean would jump ship early on the voyage to working out whether properties exist, the Carnapian has reason to stay on board, at least until he can determine whether the theories under consideration are equals as far as their philosophical utility goes, or whether one is preferable to another. We can only determine that theories are ontologically equivalent, and thus perhaps whether they are terminological variants, when we have done the metaphysical work to formulate the theories as best we can. That is, we need a project like that in Chapters 2–4 in order to realize that different metaphysical theories are equals in some important sense. There is still a reason to worry about whether properties exist from a Carnapian point of view.

9.3 Two principles (again)

What reasons can be given in support of APP or SPP? In the absence of a deductive argument, we might try to rely upon a weaker form of justification. For the case in favour, the list of applications of properties from the beginning of 9.1 could be presented as evidence to infer the objective existence of properties by inference to the best explanation: realism about properties or qualities is the best explanation of our common sense and scientific experience of classification and generalization. Note that while the ontological realist is concerned with the truth of SPP or APP, those sympathetic with Carnap's views will be answering a slightly different question: Why should we prefer a theory containing an assumption about the objective existence of properties over a theory which does not contain such an assumption? But in the course of providing justification to draw an inference to the best expla-nation, the kinds of evidence which they provide – and often the evidence itself – will be the same.

There are general difficulties with this mode of reasoning. The question of what counts as explanation, never mind what counts as the best explanation, is open to enthusiastic debate and it will be difficult to reach a general consensus. Nevertheless, the property realist might be able to establish that explanations of our classificatory and predictive practices which postulate mind-independent properties are more plausible than those which do without, and that puts his inference to the best explanation, in a relatively stronger position.

The explanatory advantages claimed for APP and SPP are likely to be different, since APP includes those properties which we encounter in our ordinary everyday interaction with middle-sized objects such as tables, trees, cats and cars, while SPP postulates the existence of a minimal set of fundamental properties which determine the qualitative nature of the rest of the world. Because of this, the former is more likely to be susceptible to support from common-sense observations of the objects and events around us and the qualities which they have. On the other hand, SPP involves more sophisticated metaphysical claims about the nature of the world,[8] since it is unlikely that we encounter sparse, fundamental properties in our everyday lives, nor perhaps do we even encounter them as a result of our best empirical science.

While the evidence for APP is plentiful and available to anyone who cares to look, SPP is another matter, so it is not obvious why one would prefer the latter to the former. But the explanatory asymmetry which initially appears to count in favour of APP is countered by concerns about the relative abundance of properties which APP would bring. Although abundance has advantages – in providing a theory of meaning, for example – it runs counter to our theoretical preference for simplicity and parsimony. APP brings with it the prospect of accepting an account of causation which allows for widespread causal overdetermination in the world; that is, the prevalence of effects which have more than one sufficient cause, such as the man who is killed by being shot, poisoned, stabbed, and struck by lightning. Or else, it leaves higher-level causally efficacious properties without underlying physical causes and thereby threatens the presumption that physics is causally complete. For those with inclinations towards physicalism, it seems peculiar that properties such as *considering a move to Memphis*, or *being the third chocolate ice cream sundae ordered for breakfast* should be entirely causally autonomous of the physical, even if they are not in a strict lawlike relation with physical properties. One might suggest that the proliferation permitted by APP is more plausible if treated as being part of our *explanatory* ontology: the abundant properties which are postulated by APP are those which appear in causal explanations and include many ways of classifying particulars which are convenient or easier to understand, but which do not obviously reflect mind-independent divisions in the world. The supporter of SPP can maintain that his principle is more parsimonious and, combined with a mechanism by which sparse properties can generate the rest, presents a more elegant account of the properties and causal interactions of the world: the metaphysical assumptions required to support versions of the SPP are worth making. I will not rehearse further arguments for SPP versus APP here (see 5.3) except to note that despite its prima facie empirical advantage, APP is probably the least metaphysically satisfactory of the two.

Just as Quine was sceptical about properties in general, some philosophers are sceptical about the entities postulated by the Sparse Properties Principle, considering them counterintuitive, or mysterious (Putnam 1981; Taylor 1993; Elgin 1995). But SPP cannot be dismissed purely because it involves significant assumptions that nature is ultimately simple, parsimonious, and most probably unified. Every metaphysical theory requires some assumptions, and the existence of something or other, and the sparse property theorist is no better nor worse off in this regard. As long as his principle can do the explanatory work which it promises, then SPP is acceptable as an assumption about which entities exist.

9.4 Do we know which sparse properties there are?

Let us allow that the Sparse Properties Principle is true, or that we have good reason to accept one of the wide range of metaphysical theories which include ontological commitment to mind-independently existing sparse properties. Now we are faced with the second question: How do we know, or have good reason to believe, *which* properties there are? Sparse properties are (most probably) not the ones which we encounter in our everyday dealings with ordinary middle-sized objects, although we may encounter some properties that way. For instance, some numerical or quantitative properties such as *being more than* or *being less than*, *being three*, or *being the third* are instantiated by objects of which we have direct sensory awareness, and so if these are among the sparse properties, we could find out about them by measuring, or counting objects or other properties. However, it seems plausible to think that most sparse properties will only be discovered by empirical investigation. If physicalism is true, then we will find the objective joints in nature by doing physics; while, if physicalism is not true, investigation in the other sciences, and perhaps other areas of research, will be required as well.

This tale of empirical discovery is the story which many sparse property theorists like to tell, or else to implicitly assume: if we investigate the world empirically, we will eventually discover which sparse properties there are to a reasonable level of determinacy. There is room for a little underdetermination or error in what we discover: we need not require that our empirical investigations will uncover the exact set of sparse properties, since that claim would be implausible, but the end result will be 'close enough' to the true nature of the world, at least insofar as the discovery of the actually instantiated sparse properties is concerned. So, for those who accept the Principle of Instantiation, we will discover all sparse properties this way (as long as there are no fundamental properties which are causally inert and thus have no effect on us or upon other things).

However, there are some arguments which suggest that this account is too optimistic: if sparse properties do exist, we may not be able to discover which they are to within a tolerable degree of determinacy. If this conclusion holds, then the case for accepting SPP on the basis of inference to the best explanation is weakened: if we lack epistemic access to the entities it postulates, it is not clear how these entities could be relevant to the explanations which we formulate, with which we predict and explain the phenomena we encounter. At least some of the purported explanatory power of SPP will have turned out to be illusory, although it may still be an acceptable claim about the existence of properties from a purely metaphysical point of view.

There are two species of sceptical arguments presented about our epistemic access to sparse properties. The first apply only to quiddistic sparse properties, in which properties are essentially qualitative and do not have their respective causal powers necessarily. The second group of arguments also applies to properties which have their causal roles necessarily.

9.4.1 Humility and problems with quiddities

We have already encountered the sceptical arguments concerning quiddities in 7.2.2: Duplication, Replacement and Permutation. If properties are quiddistic, such that they have a qualitative intrinsic nature and their causal role can change in different possible situations, then the following sceptical scenarios could arise: distinct properties could have the same effects as each other in the actual world; or, an alien property could play a familiar causal role (*mass* could be replaced by *schmass*, for example); or, the causal roles of properties could swap with each other. The epistemic problems arise because we find out about properties from their effects, and so if the effect of a property can vary this is not a reliable way to discover which properties there are.

A potential counterexample to the sceptical attack is that we might be aware of qualia directly. These are conscious properties, such as how the blue of IKB253 appears to me, or how chocolate tastes to you, which would be exceptions if we were able to detect their qualitative nature by acquaintance without recourse to their causal role. Many philosophers would reject the existence of such properties (Lewis 2009: 217–18). But even if qualia do exist, they will only provide a general solution to the sceptical problem if they are fundamental (and exhaust the sparse properties), or they are related to non-phenomenal properties in lawlike ways. However, such lawlike connections would have to hold with metaphysical necessity to circumvent the sceptical arguments, which laws holding between categorical properties are unlikely to do[9] (8.4). Thus, even the existence of qualia would not relieve the sceptical challenge to our knowledge of non-phenomenal qualitative properties of the external non-conscious world.

There are three responses to these problems with quiddities: humility; making an anti-sceptical argument; or rejection of quiddities. The first response is adopted in the later work of Lewis (2009) among others who counsels *humility* towards the sceptical conclusions of the Replacement and Permutation Arguments. We are not in an epistemic situation to discover the true qualitative nature of the world and we should simply accept this fact about our epistemic relationship to properties. The problem is worse than simply being sceptical about the sparse fundamental properties themselves, as Lewis notes, since other properties – on his view the *natural* properties – are taken to be combinations of the sparse properties which we cannot know about. The sceptical problem spreads to non-fundamental properties too, since we know neither how these properties are generated from the sparse ones, nor which sparse properties they are generated from.

The second, anti-sceptical approach responds that the scepticism involved is no more problematic than the usual, common-or-garden scepticism about the external world which we are used to, and that any theory of knowledge which permits us to have knowledge despite this broader scepticism will also deal adequately with the arguments that quiddities are not within our epistemic grasp (Schaffer 2005: 16–24). However, there are difficulties with the suggestion that standard responses to scepticism are sufficient for the task at hand. It is one thing to say that I should not be sceptical about whether there is a desk in front of me, and quite another to say that our sceptical worries should be set aside when we are concerned about discovering which quiddistic sparse properties there are. The latter, underwritten by a quiddistic version of SPP, is a sophisticated metaphysical thesis about the existence and nature of the qualitative divisions in the universe; its denial is neither counter-intuitive, nor implausible from the point of view of our ordinary experience of the world. Furthermore, the epistemic status of SPP is different to that of common-sense claims such as 'there is a desk in front of me'. Faced with an enthusiastic Cartesian sceptic, I might accept that there is a desk in front of me on the basis of inference to the best explanation: presupposing the existence of the external world and objects in it with which I interact is a better explanation of my experience of persistence and change than fanciful alternatives such as my being a brain in a vat. The claim which I thought I knew on the basis of common sense and my causal interactions with the world is reinstated by inference to the best explanation. But the case of SPP is disanalogous: we have *already* accepted a quiddistic version of the SPP on the basis of inference to the best explanation, and now the sceptical arguments against quiddities have found that SPP does not actually have the explanatory power promised; so, to invoke inference to the best explanation for a second time, in order to shore up the explanatory power of SPP which is now threatened by the sceptic, seems unprincipled to say the least.[10] With justificatory practices like that we could sustain almost any ontology we like.

We should note briefly here that APP is in a better epistemic position with respect to the sceptical problems: unlike in the case of SPP, our encounters with abundant properties are not restricted to their being the goal of scientific investigations, but are also part of our folk metaphysics. Thus, our initial acceptance of specific properties, such as *being a dog*, *being a house*, or *being blue*, is in place as part of our common-sense understanding of the world and plays a role in the initial inference to the best explanation in favour of APP. (Justified belief in such properties is on a par with my claim that there is a desk in front of me in the previous paragraph.) The same is not true in the case of SPP where the discovery of sparse properties comes later as a result of empirical endeavour and, to a great extent, because we already believe that the sparse properties are there to be discovered.

The difficulties which the supporter of the quiddistic version of SPP encounters in dealing with the sceptical arguments might be taken to recommend the third course of action: why not, as was suggested by the very same arguments in 7.2.2, take such epistemic problems to count against quiddities themselves? If we accept a dispositionalist conception of properties according to which properties have their causal roles necessarily, the arguments from Duplication, Replacement and Permutation cannot get off the ground. That, it is argued, will solve the problem posed for SPP about how we know which sparse properties there are (and thus how we know which properties there are in general) because we can discover properties by detecting their effects.

9.4.2 Causal role, laws and more permutations

The victory of the dispositionalist in the debate with the sceptic is short-lived, however: treating the causal power of a property as necessary to it simply shifts the sceptical problem and does not remove it entirely. Recall that the difference between the quiddistic account of properties and the dispositionalist account also led to differences in the ontological account of causal laws: if properties have their causal powers contingently, the natural laws which there are in each possible world direct or govern what properties can do in that world, while on the dispositionalist view, the causal powers of the properties determine which laws there are. Because causal powers belong to dispositional properties necessarily, laws and properties go together (one might think that one category reduces to another, although that feature is not important here); properties and laws are ontologically interdependent.

The sceptical problem for the dispositionalist exploits this close connection between properties and laws, since this interdependence also appears at a theoretical level. To see the problem, we have to distinguish between the predicates of our theory on the one hand and the properties which we hope to pick out with them on the other. Part of the goal of science is to discover which properties there are in the world, as well as which laws are true, and to do so we formulate generalizations and predictions, and confirm them by induction. Eventually, the story goes, we will have a complete theory which covers the fundamental causal interactions of the world, and that theory will utilise a set of *primitive* predicates which refer to sparse properties within a tolerable margin of error. But, given the inter-relations of properties and laws, this story is more complicated: to confirm laws, we have to observe regularities between types of things (where 'thing' is used loosely); but we confirm whether these things are the types that we think that they are (that they have the properties which we think they have) in virtue of their participating in certain causal generalizations. We confirm the attribution of properties and which causal laws there are on the basis of the same observations and this 'boot-strapping' – while effective for formulating a theory which predicts and explains – opens the door to a new breed of permutation argument. The difficulty is that for every theory we could formulate and confirm, there are permutations of that theory which could also be confirmed by the same observations (Davidson 1966) and yet the predicates within these permutations pick out very different properties to our favoured theory. This can be illustrated by some rather strange examples.[11]

For instance, consider the chemical terms 'silicon', 'oxygen', and 'carbon'. If we add *carbon* to *silicon dioxide* (a compound of silicon and oxygen commonly known as 'sand'), the result is pure *silicon* and *carbon monoxide* (a compound of carbon and oxygen).

Chem: Carbon + (Silicon and Oxygen) → Silicon + (Carbon and Oxygen)

Now consider the language of Strange Chemistry, which includes the terms 'siligen' and 'oxybon', where 'siligen' is equivalent to 'silicon or oxygen' and 'oxybon' is equivalent to 'carbon or oxygen'. This makes the English chemical terms equivalent as follows: 'silicon' is 'siligen which is not oxybon', 'carbon' is 'oxybon which is not siligen' and 'oxygen' is 'oxybon or siligen'.

The Strange Chemists would observe the following reaction: *Siligen* added to *oxybon which is not siligen* will result in *siligen which is not oxybon* and pure *oxybon*.

StrangeChem: Siligen + (Oxybon which is not siligen) → (Siligen which is not Oxybon) + Oxybon

While chemists think that they are purifying silicon, Strange Chemists think that they are purifying oxybon; the predicates of the two theories carve the world up differently. Given such examples, it seems that there is no guarantee that confirming theoretical hypotheses, and using them to predict and explain, will yield predicates which refer to the sparse properties which there are. The familiar theory and the Strange theory are empirically equivalent.

One might think that this conclusion has not yet taken into account other aspects of theoretical development: perhaps one could argue that (Chem) is simpler than (StrangeChem) and thus would be preferable. There are two points to note here; first, although the chemical hypothesis is simpler, Strange Chemistry is more parsimonious since it involves fewer chemical kinds; and secondly, a theory's being simple only counts in favour of its being more apt to pick out sparse properties if the objective qualitative divisions in the world are simple too. Even if they are, it does not seem obvious why our conception of simplicity would match up to the simplicity of the causal structure of the objective world unless we make what amounts to a rationalist assumption that the two coincide. The entities which SPP countenances seem to be epistemically elusive even on a dispositionalist conception of properties.

These examples are not intended to suggest that we reject the theories which we do use, unless we find empirical flaws in them. Rather, the difficulty is parallel to that of the quiddistic version of SPP in that sparse properties, and the less fundamental properties which are determined by them, may not be the ones with which we understand the world, and there is no way of finding out whether this is the case.

Moreover, even if we accept APP, there will still be a problem. Although, according to APP, we could allow that there are sufficient abundant properties for each gerrymandered predicate in every strange theory which can be confirmed, we are now faced with having multiple properties and multiple generalizations for each, intuitively single, case of causation. APP can cope with the multiplicity of permutations, but it offends against Ockham's razor to think that what happens in the world occurs in virtue of all these properties at once: surely the natural world is more efficient than that. However, the supporter of abundant properties has no way to attach relative importance to the properties picked out by one set of theoretical predicates rather than another, except insofar as we are using one set in our theories rather than the other. Thus, there are only pragmatic reasons to rule out the ontological multiplicity implied by the existence of multiple empirically equivalent theories and the over-determination of causal structure which that suggests.[12]

One might object to the Strange Chemistry example on the grounds that the Strange predicates are defined in terms of our familiar ones and that therefore the 'competing' sciences are only terminological variants of each other (just as one might argue that seemingly competing explanatorily equivalent metaphysical theories are terminological variants (Hirsch 2002, 2009)). That is, the example is either illegitimate because the Strange Chemistry predicates are not strange enough, or irrelevant even if they are. Has the sceptic over-simplified the process by which we refer to sparse properties? First, the fact that the gerrymandered predicates are defined in terms of predicates which we do use does not help the property realist, since it would be begging the question to claim that the predicates of our science are those which pick out the sparse properties there actually are. From the perspective of Strange Scientists, our predicates are defined in terms of theirs.[13] To make this response work, we need a plausible reason why the human-oriented sensory perspective from which we began our scientific theorizing yields reference to sparse properties, or properties closely enough dependent upon them that we might be able to work out how the properties which we do perceive are determined by the sparse ones. Perhaps some properties such as colours are 'response-dependent' in that we cannot help but see those properties as they actually are (Pettit 1998). However, it seems a tall order to justify the claim that any properties are response-dependent in the requisite way. Why should we think that any properties which humans are disposed to pick out, rather than those favoured by other species – real or hypothetical such as the Strange Scientists – are those which capture the qualitative divisions of the world as they actually are? Furthermore, even if such properties exist, it is not clear that they help the sparse property theorist because we do not know how these properties relate to sparse properties (presuming that response-dependent properties are non-fundamental). The Strange Science examples are not obviously illegitimate.

Second, given the inter-definition involved in the permutation of the predicates, could we object that our science and Strange Science are terminological variants of each other and thus, that the examples are unproblematic? The first point to note is that this is a legitimate response in the face of the examples for some philosophers; the second point is that it is probably not a legitimate response for property theorists who care that we know (to within a reasonable degree of determinacy) which sparse properties there are. If we take the view that we and the Strange Chemists are both right and yet we do not wish to accept that the predicates of the two sciences pick out different properties as they appear to do, then we have given up on aiming for a direct one-to-one correlation between theoretical terms and objectively existing sparse properties. There is, on this view, some kind of 'magnetic' referential attraction by which theoretical terms latch on to sparse properties, but if we and the Strange Chemists can refer equally well, it is not obvious that this reference relation does anything more than stipulate its way out of the sceptical problem, nor that we have much understanding of which properties there are as a result of it. After all, the competing theories appear to carve the world up in different ways and this response amounts to asserting that they do not; so, the claim that our science will tell us about the qualitative divisions in nature seems impossible to uphold. On the other hand, if there is an asymmetry of reference between our case and that of the Strange Scientists – that is, if we should prefer the predicates of one theory over another – one would have to provide a reason as to why this would be and will face similar problems to those encountered in the previous paragraph.

Even if we accept the examples, there are still ways in which we might attempt to rescue our knowledge of mind-independent properties from the sceptical problem above. First, as in the case of the quiddistic version of SPP, we might make an anti-sceptical move and say that the problem of Strange Science is no different to other, more familiar sceptical problems and so this version of scepticism should not trouble us unduly. However, as when this strategy was used to deal with the quiddistic versions of SPP, there is a disanalogy between disarming scepticism about our common sense claims about the world and disarming scepticism about a specific and sophisticated metaphysical assumption. As above, this would be a case of using inference to the best explanation to shore up a claim which we only take to be true by inference to the best explanation in the first place, and that is philosophically tendentious. Philosophers are entitled to make assumptions – in fact, they could not do without them – but the plausibility of a theory decreases if additional assumptions are required to save it from counterexamples. On the other hand, the sceptical argument, despite the strangeness of its predicates, plays on quite a common-sense observation: we are happy to accept that the terms of different languages carve the world up in different ways (sometimes in ways which we struggle to reconcile) each providing roughly equivalent information in the form of truths about the world to its speakers. What the Strange Science examples do is to imagine a similar situation with respect to a scientific theory; it is not too implausible to think that scientists working in isolation from each other would divide the world up in different ways, barring any clear mechanism by which they would all detect the same sparse properties. The usual response to scepticism does not seem appropriate in this case.

The second strategy for responding to the sceptical problem is to question whether the argument has thoroughly explored all the ways in which we might find out about mind-independent properties and how the predicates of our theories refer to them. This project coincides with the defence of Scientific Realism in the philosophy of science, which attempts to uphold the claim that our best scientific theories are objectively true, or that they tell us which entities the natural world contains to within a tolerable margin of error. Perhaps our theory and all the Strange Science permutations will converge to one at a fundamental level; or, maybe one can argue that science would not work for prediction and explanation if it did not pick out the entities which objectively exist in the world (Putnam 1975c: 73). One of the primary challenges for this family of responses is to give an account of how the predicates of our theories refer to mind-independent properties, or the generalizations we have confirmed capture the natural causal and structural order, in the light of theory change over time, or permutation problems like that of Strange Science. There is a lot to be said about whether a version of scientific realism might succeed, but due to considerations of space, I will be forced to set this project aside (see Allen 2002: 534–9 and further reading). Instead, I will explore the implications for property theorists if this response fails. After all, if it does not fail, we can go back to talking about, and dealing with, the sparse objectively existing properties discussed in this book as if the sceptical challenges had never occurred.

9.5 The implications of scepticism about properties

The tentative conclusion reached about our epistemic position with respect to mind-independent properties is that while it is legitimate from a metaphysical point of view to presuppose that there are some, we are then faced with serious difficulties in discovering which ones there are, whether these are quiddistic properties or dispositional ones. This sceptical problem may be taken to count against accepting SPP or APP in the first place, since sparse properties were accepted on the basis of their explanatory power and that has been found to be wanting in one important respect. Nevertheless, for those who accept realism – that is, they are prepared to accept that some truths may be outside our epistemic grasp – there are still good metaphysical reasons for accepting either the quiddistic or the dispositional version of SPP. Then, one can adopt a humble attitude towards our knowledge of the qualitative distinctions in the world (Lewis 2009).

However, given the problems concerning our epistemic access to sparse properties, one might suggest that we should also take an interest in whatever it is we *are* talking about when we talk about properties and use them in our best scientific theories. Whatever those entities are, it seems likely now that they are not the mind-independent sparse properties which we have previously presumed them to be. So what are they? This question is of interest whether or not one accepts the existence of an ontology of mind-independent properties alongside such entities or ultimately rejects it; and I will explore this question in 9.6.

If the sceptic is right, our theories do not refer to the objective qualitative divisions in the world or, if they do, that is purely a coincidence which we have no way to justify. So, the second implication of scepticism about properties is that it is not obvious whether we should take philosophical questions which arise because of features of our scientific worldview to have ontological import. For instance, the philosophical problem of mental causation, which arises from the competition for causal efficacy between physical properties and mental ones, might be attributable to features of our theories and how we develop them, rather than being a deep ontological problem about the causal efficacy of the mind. Similar considerations apply to discussions of reduction in general, where one family of properties is found to be determined by another, either by the discovery of 'bridge laws' which correlate the properties of two different theories, or because we can give a functional analysis of the properties of one theory in terms of another (Nagel 1961; Hempel 1965; Fodor 1974; Kim 1998). Instead of the success of such reduction having ontological implications – that there really are fewer mind-independent properties in the world than we thought – we could return to treating successful reduction as a theoretical triumph which produces a neater and more unified theory with increased explanatory power.[14]

9.6 What are properties if they are not the joints in nature?

If objective qualitative joints in nature do not exist, or are epistemically elusive, we need an alternative, or an additional, account of properties to fit our theoretical practices. The question of what our theoretical ontology amounts to if the divisions within it do not mark objective joints in nature has been touched upon by several philosophers but not, as far as I am aware, given a thorough formulation (Taylor 1993; Elgin 1995). There are (at least) two reasons that this omission might be acceptable: first, one might think that what there are instead of, or in addition to, objective qualities are predicates, ways of describing, classifying and generalizing about our experience, which are essentially linguistic entities. Second, if there are properties which are not mind-independent, one might think that one does not have to give a 'separate' theory of what such properties are, over and above one or other of the various formulations of property theory given in the first portion of this book. The reason for this is that those accounts of properties did not rely upon the objectivity of properties as such, and some of them may turn out to be plausible even if properties (and at least some of the other entities postulated by the respective ontological theories) are not objectively existing aspects of a mind-independent world. (It also may turn out that some of the ontological accounts of properties are implausible if they are not dealing in objectively existing entities and are thereby ruled out.)

The first option, which accounts for classification purely in terms of the application of predicates, is not prima facie plausible. For a start, there seem to be too many or too few predicates to do the work of the properties we need, especially if we are trying to develop a theory of sparse properties of the kind we might use in science to understand the causal structure of the world. There are too many predicates because many, or most, predicates are redundant when it comes to the task of classifying scientifically relevant types or kinds; while there are too few because we do not yet have predicates for all the sparse properties, some important predicates appear to be missing. We could deal with the first problem by introducing a less egalitarian account of predicates: perhaps not all pieces of descriptive language are equal and the predicates which are more central to our best theories are more primitive or important than the rest. But the second difficulty – of crucial predicates being missing from our best explanations – is more difficult to solve with a purely linguistic ontology of predicates. If we do not have sufficient predicates to explain all the causal interactions of the world which we observe, one might be tempted to ask what makes it the case that there is something missing from our theoretical classification system of predicates. We can develop or formulate new predicates; but *why* would we do that? The most tempting answer to that question is that there is something that we have not yet described, or described adequately, or a type of thing which we have not yet named. However, these answers suggest that there are features of the world whose effects we can detect for which we do not yet have predicates: it seems that properties are sneaking back into the ontological picture. This argument does not imply that the properties concerned exist mind-independently (although they may do) because it is precisely because we are aware that there is something which our linguistic resources are not adequate to describe that we believe that there are too few predicates. Thus, without further reason to think that the sceptical arguments fail (which we are assuming that we do not have), this falls short of being an argument for the existence of determinate qualitative joints in nature. What it does imply however is that language alone is not enough to provide the ontology of properties, even if we deny that they are discrete mind-independent entities: there seems to be a qualitative phenomenon which drives the development and improvement of language and theory via its causal action in the world.

This account does not collapse back into a version of realism about properties, but into the second view that properties may be construed as mind- or theory-dependent entities. On this view, the qualitative ontology is not entirely objective, but nor is it entirely subjective or intersubjective either; that is, it is not entirely dependent upon our perspective on the world, nor entirely independent either. What we might say is that qualities exist, or something qualitative or causally powerful exists mind-independently, but the distinctions or divisions between one property and the next are determined by our theory or theories. There are some areas of theorizing where we already accept such a view of properties. For instance, in the case of colours, different languages carve up the visible spectrum in very different ways, ranging from the two, or four colours to those with seven, eight or more; moreover the distinctions between these colours are not always drawn in the same place. Similarly, the divisions between biological species can be drawn in different ways depending upon whether we distinguish them on the basis of observable similarities and differences, interbreeding potential, genetic make-up and so on, and so whether two animals count as belonging to the same species or to different ones can only be answered relative to the other distinctions which the theory makes. A third example of theoretically-driven distinctions between properties is that drawn on the periodic table between different elements, which are then subdivided into isotopes in order to capture what can interact with what. It would make as much sense to regard this dependency as running in the other direction with the isotopes marking fundamental differences between types which can then be grouped into elements, since this classification is perhaps better at determining which chemical reactions each atom can enter into.

If the sceptical arguments of the previous sections are correct, then one might suggest that the nature of properties which is suggested by these examples can be generalized to other properties as well, those in theories where we do not have ready empirically equivalent alternatives except for gerrymandered Strange Science examples. These partially theoretical or instrumental entities may coexist alongside mind-independent properties, or else be imposed upon, or help to arrange, a mind-independent reality which is utterly unlike that suggested by our theories.

This agnosticism about what exists mind-independently might also help to avert a potential objection to the view that properties are (at least partially) dependent upon our theories: Why does the world appear to be orderly if the order is imposed by us? More generally, the Cause-Law Thesis, that every instance of singular causation is subsumed by a law, appears to be true (8.2).

The hard-headed response might be that we cannot help but see the world as containing qualitative or causal order even if it does not, since the way in which we understand what is around us essentially involves classifying and generalizing. We cannot step out of our language or belief system to conceive of a world in which properties are lacking (Davidson 1995). This hard-headed answer is available to both those who equate properties with predicates, and to those who espouse the view under discussion in which properties are determined by both theory and the world. However, the latter conception of theoretical properties can also accept that the appearance of qualitative and causal order might be indicative of some mind-independent process or other which appears orderly to us, although it need not consist in determinate properties which divide the world in the way that our theories do. One such contender is the account of how order appears in non-equilibrium systems due to the second law of thermodynamics: entropy decreases locally in such systems (rather than increasing as would be expected in a system in equilibrium) as structures appear which dissipate energy more effectively than in systems without structure (Nicolis and Prigogine 1977; Prigogine and Stengers 1984).

But does the non-realist view of properties fail the compulsory Moorean question to explain objective qualitative similarity and difference? Armstrong might claim that it does. However, as Taylor points out, it is not clear that an explanation of the *objectivity* of similarity and difference is required, rather than an explanation of the appearance of similarity and difference, which is a different matter entirely (1993). We can agree with Moore that there are certain common-sense facts which our theories should be required to explain, but the objectivity or lack of objectivity of the phenomena we intuitively accept should be part of the *analysis*, not included as part of what needs to be explained, for fear of begging the question about what exists mind-independently.

If properties are neither predicates, nor objectively existing qualities, then what are they? It seems likely that there will be a choice of plausible theories, just as there was when the ontological basis of properties was discussed earlier in this book. As I noted above, one might take one or other of those theories and adapt them to the currently required purpose. One such suggestion develops a broadly Lewisian account of sparse properties, but rejects the objective notion of property naturalness in favour of linking the fundamentality of properties to their theoretical importance, making sparse fundamental properties those which are the most central to our best theories. Such a view is also available to those who prefer an ontology of properties determined by natural classes of tropes, in which the classes would no longer be regarded as being entirely determined naturally or objectively, but partially in virtue of us. Alternatively, one might relocate the *resemblance* in theories which define properties in terms of resemblance classes, such that resemblances are those which we pick out, or are at least partially determined by us and the development of our theories, and so the resemblance classes are not entirely objective. Finally, it might be possible to reformulate theories of transcendent universals in terms of their being related to meanings or concepts, where these are somehow dependent upon how we understand the world.

There are two points to note about such reformulations of property theory which characterize determinate properties as less-than-objective entities: first, these theories are not inevitably vulnerable to some of the well-known objections to concept or predicate nominalism suggested by Armstrong (1978a). For instance, Armstrong objects that such theories get the direction of explanation 'the wrong way around', since they assert that particulars are of a type because a predicate or a concept applies to them, or they can be described in a certain way (rather than the predicate applying because the particular is of a certain type). However, this objection relies upon the desirability of giving an objective account of qualitative sameness, and that goal has already been rejected within these accounts of properties. Furthermore, as indicated above, these theories are not tied to egalitarianism about properties – they do not have to treat all properties equally, although earlier versions of them sometimes did so – and thus they are able to give accounts of sparse properties as well as abundant ones. What is required here is a concept analogous to naturalness which need not exist mind-independently but which can rank properties into a hierarchy such that some properties are more important or fundamental than others. With this theoretical apparatus, an account of sparse theory-dependent properties can be provided.

The second important point is that the adoption of such accounts of properties will significantly change our understanding in certain key areas of philosophy in which properties have proved useful. For instance, the metaphysical accounts of causality, laws and modality which properties have facilitated will no longer be objectively true if we accept that properties are theory-dependent entities and reject the alternative accounts of modality and the like in terms of the objective properties postulated by SPP and APP as well. For instance, it will not be literally, constitutively true that a cause has its effects in virtue of the properties which it has, although we might be able to say that a cause has an effect in virtue of something about it. This might be regarded as an advantage of this account of properties, as it may help to solve the problem of mental causation for example (Davidson 1970, 1993), although others will regard it as an unacceptable loss of explanatory power. Similarly, the change in the causal ontology will have an effect on what the ontological status of laws of nature could be: if we avoid reifying determinate, objectively individuated properties, or equivalent entities, then the laws which either link, or govern the behaviour of types of particulars will be theory-dependent or instrumental too. Third, the accounts of modality which rely upon properties or causal powers will also be affected; what is possible will at least partially be a mind- or language-dependent matter.

9.7 Conclusion

In the course of this chapter, we have investigated the assumption that properties are mind-independent entities and found that while there are good metaphysical reasons for accepting that such entities exist, there are also good epistemic reasons to suggest that we would not be able to find out which properties there are. If taken seriously, this scepticism opens up a new line of enquiry – what are we talking about when we apparently refer to properties with our predicates? – with the possibility that this will reveal a range of entities which involve qualitative or causal nature in the world, but depend upon us and our theorizing to be individuated as determinate entities. Instead of, or perhaps in addition to, the mind-independent ontology of properties which determine the causal and modal order of the objective world, we have an explanatory ontology which is an approximation from our epistemic perspective of what there is. While the formulation of such an account of theory-dependent properties could borrow much from what has been said about their mind-independent cousins, more work needs to be done to clarify it and the implications of holding such a view.

FURTHER READING

On existence:
Carnap 1947; Quine 1948, 1969; Chalmers et al. (eds), 2009.
On scepticism about properties and replies:
Taylor 1993; Elgin 1995; McGowan 2002; Allen 2002; Schaffer 2005; Lewis 2009.

Suggested Questions
1 Does it make sense to ask whether properties exist? What is the best way to justify the claim that they do?
2 Can we respond to the sceptical conclusion that we don't know which sparse properties there are in the same way as we would respond to a Cartesian radical sceptic?
3 Is it any easier to discover abundant properties than sparse ones?
4 Is our epistemic access to dispositional properties any more secure than to categorical properties? Can either account respond better to the sceptical arguments?

5   Can the Strange Science examples be neutralized by considering theoretical virtues such as parsimony and simplicity?

6   If properties do not objectively exist, what are they?

Notes

1   Elsewhere I have characterized a very similar principle as 'The Natural Properties Principle' (2002) to capture the specific assumption about the existence of natural properties in Lewis's property theory. The Natural Properties Principle thereby counts as a more restricted version of the Sparse Properties Principle.

2   Dennett (1989) and Davidson (1970) consider at least some mental properties specifically, propositional attitudes to be instrumental, or to have a different ontological status from other properties such as physical ones. (Davidson, however, is a realist about neither.)

3   The verb 'to Quine' is defined in *The Philosophical Lexicon* (Dennett 1987) as 'To deny resolutely the existence or importance of something real or significant', as used in Dennett's own 'Quining Qualia' (1988).

4   The difficulty from a Quinean perspective is that modal notions require identity criteria too; and even worse, the attempt to explicate necessity ultimately requires recourse to meanings, and then perhaps to entities such as properties which determine meanings (if one takes a less than empiricist position, with which Quine was not concerned). For Quine, modality and meaning (and thus, also perhaps properties) are interdefined (1951).

5   We can contrast the case of properties with that of mathematical or set-theoretic objects here. According to Quine, the latter are indispensable to our theories and so we are justified in thinking that some mathematical abstract objects must exist.

6   It is common to read Quine as more of a realist about ontology, but I do not think that interpretation is justified. I will not digress to discuss the correct interpretation here, but see Price 2009.

7   Strictly, in Carnap's view, external questions are meaningless. I am being careful in this discussion of the Carnapian attitude to metaphysics not to rely heavily upon his use of the analytic-synthetic distinction, a use which was criticized by Quine (1951). For Carnap himself, internal questions about the existence of entities are analytically true or false, while external questions are meaningless, since terms can only have meanings within a theory. However, even without this strict semantic division between internal and external questions, one might think that the former are in some sense trivial, while the latter are unanswerable since they require information about entities which is outside our epistemic grasp.

8   For a more detailed argument that SPP involves substantial metaphysical assumptions, see McGowan 2002.

9   Arguments for the metaphysical possibility of zombies rely upon laws
    between non-conscious and conscious properties holding with no more
    than nomological necessity (Chalmers 1995); so dispositionalists who
    consider laws to be metaphysically necessary may rule out zombies as a
    result.

10 Lewis proposed a version of this response to scepticism in his earlier work,
    by introducing a presumption that we refer to natural properties because
    they are constitutive of the content of our thought (1983a: 224). See Allen
    2002, for objections.

11 These are based on Hirsch (1993: 80) but have been complicated in order
    that Chemistry and Strange Chemistry do not overlap property terms.

12 One might argue that we can have direct epistemic access to some
    abundant properties. See below and Pettit 1998, and Miščević 1997 for
    objections.

13 A similar point is made by Goodman about 'grue' and 'bleen' (1983 [1954]:
    79).

14 One might object that explanatory power is lost with reductions because of
    the simplifications involved to make the kinds and structures of the different
    theories match up (Morrison 1994). I will not explore this problem here.

10

Conclusion

This final section of the book is optimistically titled; named at a time when I thought that there might be a definitive conclusion to write. That is not to say that there are no conclusions to be drawn from what has gone before – in fact, there are plenty – but if the preceding nine chapters of metaphysics have achieved anything, they should have shown that this is not a subject in which there are conclusive reasons for holding one position rather than another. On the subject of ontology: we need a theory of properties, but any one of several will do.

What should we make of this pluralism? On one hand, it might be regarded as cause for concern: if one thinks that there is a way that the world is, ontologically speaking, then it is dissatisfying not to have the philosophical ammunition to discover which way this is. Some readers may take their intuitions to be sufficient to sway them in the direction of one theory rather than another; while others might be pragmatically prepared to accept the explanatory equivalence of each of the outlined theories, regardless of the fact that they believe one or other of them to be correct (whether or not they know which one that is) (see Bennett 2009; Chakravartty forthcoming). I will not pause at this stage to evaluate arguments for realism about ontology and in the absence (or just ignorance) of such an argument, perhaps the plurality of property theories should be regarded as a reason for optimism: after all, having several viable theories is better than having none. Furthermore, perhaps it is not really pluralism at all and is one theory dressed up in different languages. But a good deal more argument would be required to establish that point and so I shall not pursue it here.

If the fact that there are several viable theories still alarms some philosophers, they would do well to remember that plenty of half-formed theories have fallen by the wayside: the theories which remain are at least good ones, they are consistent and (as far as I can tell) lack obvious flaws; they are roughly equal in their complexity and in the extent to which they rely upon assumptions which have to be treated as primitive. Nevertheless, one may feel that there is still a right answer to be had here – the world really is one way or another – even though we have not yet managed to determine what it is. Because, ultimately, I think that determining which theory of properties is the most plausible relies upon pre-theoretical intuitions about how the world is, the theory's consistency with one's other philosophical commitments, and a way of weighting theoretical virtues relative to one another which defies codification, it is at this point in the book that I will leave the reader to choose for him- or herself.

All this is not to say that I am unable to draw my own conclusions. Intuitively, my sympathies lie with the sceptics of Chapter 9, who are concerned that there is no way to determine whether the properties which we pick out are those which there are. Although the ontological assumption that there are qualitative joints in nature is innocuous enough, I am nervous about commitment to a category of entities when there is no good reason to think that the conception of properties which we have matches the way that the world is divided, if it is determinately divided at all. This anti-realism does not extend to sceptical concerns about the external world in general however, since I think that scepticism about properties is different in kind to radical scepticism, and that the interactions of the world still contribute to the classifications and generalizations we make.

If, on the other hand, I overlook the sceptical problems and become more sympathetic to realism about properties – and realism about metaphysics in general – I find myself swaying in the direction of a trope theory in which properties have their causal roles necessarily. Tropes have, I think, an advantage over the theory of universals, which never quite shakes off the conceptual difficulties about location associated with immanence and the questions about abstract objects associated with transcendent universals. A set-theoretic account of properties, on the other hand, seems to make more sense from a non-realist perspective given the metaphysical or modal complications required to provide an account of properties in resemblance or class nominalist terms. From my own perspective, it seems more plausible to think of the qualitative aspects of the world as fine-grained particular entities bearing resemblance relations to each other as a matter of brute fact, although perhaps such resemblance relations are not always, or are never, quite exact.

Glossary

These definitions are intended as a guide only. They are not meant to be stipulative and may be open to objection or revision.

Abstract particular Also *trope, mode, moment, way of being, event aspect, property-instance*. A particular unrepeatable quality.

Abstract particular A particular object which is also abstract (e.g. a set or class). This term is *not* used in this way in the text.

Actualism The view that what is possible or necessary is entirely determined by entities which exist in the actual world.

Bleen Blue if observed before time t, otherwise green. (t is in the future.)

Causal theory of knowledge The thesis that a subject S knowing that P essentially involves a causal process, for instance S's belief that P being caused by the fact that P.

Cause-Law Thesis The claim that every instance of singular causation is covered by a regularity or law.

Combinatorialism The view that what is possible or necessary is determined by combinations and recombinations of actual properties.

Compresence A relation (or similar) which binds spatio-temporally co-located tropes into a concrete particular.

Concrete particular A particular, unrepeatable event or object.

Counterpart A particular C_1 in world w_1 is a counterpart of another particular C_2 in w_2 if C_1 corresponds to C_2 more closely than any other particular in w_1 corresponds to C_2. (The counterpart relation replaces *transworld identity* in modal realism.)

Determinate/determinable A property is a determinate of a determinable property if the latter is an overarching category or class of which the former is a specific case. For example, *red* is a determinate of the determinable *colour*, or *1.45 kg* is a determinate of the determinable *mass*.

Epiphenomenal Lacking in causal efficacy.

Equivalence relation A relation which is symmetric, transitive and reflexive.

Event (Ambiguous) (i) a species of concrete particular (Davidson); (ii) a structured complex entity similar to a state of affairs (Kim, Lewis).

Grounding A constitutive form of ontological determination.

Grue Green if observed before time t, otherwise blue. (t is in the future.)

Haecceity This-ness, or whatever makes a particular, unique, the one particular it is.

Hyperintensional A difference between entities is hyperintensional if necessarily coextensive entities are distinct. Hyperintensionally individuated entities are more fine-grained than those individuated by possible differences between them. E.g. if A and B are coinstantiated in all possible situations and are nevertheless distinct, there is a hyperintensional difference between them.

Internal relation A relation, the existence of which is determined entirely by its relata.

Mereology The study and analysis of part-whole relations, composition, etc.

Modal realism The metaphysical view that possible worlds, and the entities in them, exist in the same sense as the actual world and its contents.

Nomological/nomic Lawlike, or pertaining to laws of nature.

Reflexive A binary relation R is reflexive iff for any element d of the domain, Rdd.

Singular causation A particular instance of cause and effect.

State of affairs A structured, complex entity consisting of a thin particular

having a property (or instantiating a universal) at a time. E.g. Particular b being F at t_1.

Substrate/substratum That which exists independently of any properties; substance in which properties are instantiated.

Supervenience Determination relation between properties or families of properties. Family A supervenes upon family B iff there can be no change in A without a change in B.

Symmetry A relation R is symmetric iff for any elements d, e of the domain: if Rde, then Red.

Thick particular A concrete particular including all its properties.

Thin particular A concrete particular minus all its properties.

Transitivity A binary relation R is transitive iff for any elements d, e, f of the domain: if Rde and Ref, then Rdf.

Underdetermination A theory T_1 is underdetermined by evidence for it E, if E does not entail T_1 rather than alternative theory T_2.

Bibliography

Achinstein, P. 1974. 'The Identity of Properties.' *American Philosophical Quarterly* 11: 257–75.

Achinstein, P. 1983. *The Nature of Explanation.* New York: Oxford University Press.

Allen, S. R. 2002. 'Deepening the Controversy over Metaphysical Realism.' *Philosophy* 77: 519–41.

Allen, S. R. 2012. 'What Matters in (Naturalized) Metaphysics?' *Essays in Philosophy* 13: 211–41.

Allen, S. R. 2015. 'Curiosity Kills the Categories: A Dilemma about Categories and Modality.' *Metaphysica* 16: 211–30.

Allen, S. R. (Manuscript.) 'Intrinsicality and Grounding.'

Anscombe, G. E. M. 1971. 'Causality and Determination.' Reprinted in Sosa and Tooley (eds), 1993: 88–104.

Armstrong, D. M. 1978a. *Nominalism and Realism. Universals and Scientific Realism* Vol. 1. Cambridge: Cambridge University Press.

Armstrong, D. M. 1978b. *A Theory of Universals. Universals and Scientific Realism.* Vol. 2. Cambridge: Cambridge University Press.

Armstrong, D. M. 1980. 'Against "Ostrich" Nominalism: A Reply to Michael Devitt.' *Pacific Philosophical Quarterly* 61. Reprinted in Mellor and Oliver (eds), 1997: 101–11.

Armstrong, D. M. 1983. *What is a Law of Nature?* Cambridge: Cambridge University Press.

Armstrong, D. M. 1988. 'Are Quantities Relations: A Reply to Bigelow and Pargetter.' *Philosophical Studies* 54: 305–16.

Armstrong, D. M. 1989a. *Universals: An Opinionated Introduction.* Boulder, CO: Westview Press. (75–112 reprinted as 'Universals as attributes' in Loux (ed.), 2001: 65–91.)

Armstrong, D. M. 1989b. *A Combinatorial Theory of Possibility.* Cambridge: Cambridge University Press.

Armstrong, D. M. 1993. 'The Identification Problem and the Inference Problem.' *Philosophy and Phenomenological Research* 53: 421–422.

Armstrong, D. M. 1997. *A World of States of Affairs.* Cambridge: Cambridge University Press.

Armstrong, D. M. 1999. 'The Causal Theory of Properties: Properties According to Shoemaker, Ellis, and Others.' *Philosophical Topics* 26: 25–37.

Armstrong, D. M. 2004. *Truth and Truthmakers.* Cambridge: Cambridge University Press.

Armstrong, D. M. 2005. 'Four Disputes about Properties.' *Synthese* 144: 309–20.

Bacon, J., Campbell, K. and Reinhardt, L. (eds) 1993. *Ontology, Causality and Mind.* Cambridge: Cambridge University Press.

Banai, N. 2004. 'From the Myth of Objecthood to the Order of Space: Yves Klein's Adventures into the Void.' In O. Berggruen (ed.), *Yves Klein*, exhibition catalogue, Schirn Kunsthalle Frankfurt, Frankfurt.

Barrett, R. and Gibson, Roger F. (eds) 1990. *Perspectives on Quine*. Oxford: Blackwell.

Bauer, William A. 2011. 'An Argument for the Extrinsic Grounding of Mass.' *Erkenntnis* 74: 81–99.

Beebee, H. 2000. 'The Nongoverning Conception of Laws of Nature.' *Philosophy and Phenomenological Research* 61: 571–94.

Benacerraf, Paul. 1973. 'Mathematical Knowledge.' *Journal of Philosophy* 70: 661–80.

Bennett, Karen. 2003. 'Why the Exclusion Problem seems Intractable and how (just maybe) to Tract it.' *Noûs* 37: 471–97.

Bennett, Karen. 2009. 'Composition, Colocation and Metaontology.' In Chalmers et al. (eds), 2009: 38–76.

Bennett, Karen. (Forthcoming.) *Making Things Up*. Oxford: Oxford University Press.

Benovsky, J. 2014. 'Tropes or Universals: How (not) to Make One's Choice.' *Metaphilosophy* 45: 69–84.

Bird, A. 2007. *Nature's Metaphysics*. Oxford: Oxford University Press.

Black, R. 2000. 'Against Quidditism.' *Australasian Journal of Philosophy* 78: 87–104.

Borghini, A. and Williams, N. E. 2008. 'A Dispositional Theory of Possibility.' *Dialectica* 62: 21–41.

Bottani, A., Carrara, M., Giaretti, P. (eds) 2002. *Individuals, Essence and Identity*. Dordrecht: Springer Verlag.

Braddon-Mitchell, D. and Nolan, R. (eds) 2009. *Conceptual Analysis and Philosophical Naturalism*. Boston, MA: MIT Press.

Bradley, F. H. 1893. *Appearance and Reality*. London: Swan Sonnenschein.

Broad, C. D. 1933. *Examination of McTaggart's Philosophy: Vol. 1*. Cambridge: Cambridge University Press.

Burge, Tyler. 1979. 'Individualism and the Mental.' *Midwest Studies in Philosophy* 4: 73–121.

Campbell, Keith. 1981. 'The Metaphysic of Abstract Particulars.' *Midwest Studies in Philosophy VI: The Foundations of Analytic Philosophy*. Reprinted in Mellor and Oliver (eds), 1997: 125–39.

Campbell, Keith. 1990. *Abstract Particulars*. Oxford: Basil Blackwell.

Carnap, Rudolf. 1947. *Meaning and Necessity*. Chicago: Chicago University Press.

Cartwright, Nancy. 1980. 'The Reality of Causes in a World of Instrumental Laws.' In Cartwright, 1983: 74–86.

Cartwright, Nancy. 1983. *How the Laws of Physics Lie*. Oxford: Clarendon Press.

Cartwright, Nancy. 1989. *Nature's Capacities and their Measurement*. Oxford: Oxford University Press.

Cartwright, Nancy. 1999. *The Dappled World: A Study of the Boundaries of Science*. Cambridge: Cambridge University Press.

Chakravartty, Anjan. (Forthcoming.) 'Particles, Causation, and the Metaphysics of Structure.' *Synthese*.

Chalmers, A. 1999. 'Making Sense of Laws of Physics.' In Sankey (ed.), 1999: 3–16.

Chalmers, David. 1995. 'Facing up to the Problem of Consciousness.' *Journal of Consciousness Studies* 2: 200–19.

Chalmers, David. 2009. 'Ontological Anti-realism.' In Chalmers et al. (eds), 2009: 77–129.

Chalmers, David J., Manley, D. and Wasserman, R. (eds) 2009. *Metametaphysics: New Essays on the Foundations of Ontology.* Oxford: Oxford University Press.

Chisholm, R. (ed.) 1960. *Realism and the Background of Phenomenology.* Glencoe, IL: Free Press.

Cohen, S. M. 1971. 'The Logic of the Third Man.' *Philosophical Review* 80: 448–75.

Correia, F. and Schnieder, B. (eds) 2012. *Metaphysical Grounding: Understanding the Structure of Reality.* Cambridge: Cambridge University Press.

Cumpa, J. 2012. '"In One": The Bearer Issue and the Principles of Exemplification.' *Axiomathes*: 201–11.

Daly, Chris. 1997. 'Tropes.' In Mellor and Oliver (eds), 1997: 140–59.

Davidson, Donald. 1966. 'Emeroses by Other Names.' *Journal of Philosophy* 63: 778–9. Reprinted in Davidson, 1980: 225–27.

Davidson, Donald. 1967. 'Causal Relations.' Reprinted in Davidson, 1980: 149–62.

Davidson, Donald. 1970. 'Mental Events.' Reprinted in Davidson, 1980: 207–25.

Davidson, Donald. 1980. *Essays on Actions and Events.* Oxford: Clarendon Press.

Davidson, Donald. 1993. 'Thinking Causes.' In Heil and Mele (eds), 1993: 3–17.

Davidson, Donald. 1995. 'Laws and Cause.' *Dialectica* 49: 263–79.

Dennett, D. 1987. *The Philosophical Lexicon.* Oxford: Blackwell.

Dennett, D. 1988. 'Quining Qualia.' In Marcel and Bisiach (eds), 1988.

Dennett, D. 1989. *The Intentional Stance.* Cambridge, MA: MIT Press.

Divers, J. 2002. *Possible Worlds.* London: Routledge.

Dretske, F. 1977. 'Laws of Nature.' *Philosophy of Science* 44: 248–68.

Ducasse, C. J. 1926. 'On the Nature and Observability of the Causal Relation.' Reprinted in Sosa and Tooley (eds), 1993: 125–36.

Dupré, John. 1993. *The Disorder of Things.* Cambridge, MA: Harvard University Press.

Dupré, John. 2012. *Processes of Life: Essays in the Philosophy of Biology.* Oxford: Oxford University Press.

Eberle, R. 1975. 'A Construction of Quality Classes Improved upon the *Aufbau.*' In Hintikka (ed.), 1975: 55–73.

Ehring, Douglas. 1997. *Causation and Persistence.* Oxford: Oxford University Press.

Ehring, Douglas. 2011. *Tropes: Properties, Objects and Mental Causation.* Oxford: Oxford University Press.

Elgin, Catherine Z. 1995. 'Unnatural Science.' *Journal of Philosophy* 92: 289–302.

Ellis, B. 2000. 'Causal Laws and Singular Causation.' *Philosophy and Phenomenological Research* 61: 329–51.

Ellis, B. 2001. *Scientific Essentialism.* Cambridge: Cambridge University Press.

Figdor, Carrie. 2008. 'Intrinsically/Extrinsically.' *Journal of Philosophy* 105: 691–718.

Fine, Kit. 1994. 'Essence and Modality.' *Philosophical Perspectives* 8: 1–16.
Fine, Kit. 2002. 'The Varieties of Necessity.' In Gendler and Hawthorne (eds),
    2002: 253–81.
Fine, Kit. 2012. 'A Guide to Ground.' In Correia, F. and Schnieder, B (eds), 2012:
    37–80.
Fodor, Jerry. 1974. 'Special Sciences.' *Synthese* 28: 97–115.
Fodor, Jerry. 1990. 'Making Mind Matter More.' In *A Theory of Content*,
    Massachusetts: MIT Press: 137–60.
Francescotti, Robert. 1999. 'How to Define Intrinsic Properties.' *Noûs* 33:
    590–609.
Frege, Gottlob. 1884. *Die Grundlagen der Arithmetik*. Translated by J. L. Austin
    (1950) as *The Foundations of Arithmetic*. Oxford: Blackwell.
Gaskin, R. 2008. *The Unity of the Proposition*. Oxford: Oxford University Press.
Gendler, S. and Hawthorne, J. (eds) 2002. *Conceivability and Possibility*. Oxford:
    Oxford University Press.
Giberman, Dan. (Manuscript.) 'Discerning Universals from Particulars.'
Goodman, Nelson. 1972. 'Seven Strictures on Similarity.' In his *Problems and
    Projects*. Indianapolis: Bobs-Merrill: 22–32.
Goodman, Nelson. 1977 [1951]. *The Structure of Appearance*. Dordrecht:
    D. Reidel.
Goodman, Nelson. 1983 [1954]. *Fact, Fiction and Forecast*. Cambridge, MA:
    Harvard University Press.
Granger, G. G. 1983. 'Le problème de la "Construction Logique du Monde".'
    Revue Internationale de Philosophie 37: 5–36.
Groff, R. and Greco, G. (eds) 2012. *Powers and Capacities in Philosophy: The
    New Aristotelianism*. London: Routledge.
Hale, B. and Hoffmann, A. (eds) 2010. *Modality: Metaphysics, Logic and
    Epistemology*. Oxford: Oxford University Press.
Harris, R. 2010. 'How to Define Extrinsic Properties.' *Axiomathes* 20: 461–78.
Hausman, A. 1979. 'Goodman's Perfect Communities.' *Synthese* 41: 185–237.
Hawthorne, J. 2001. 'Intrinsic Properties and Natural Relations.' *Philosophy and
    Phenomenological Research* 63: 399–403.
Heil, John. 2003. *From an Ontological Point of View*. Oxford: Oxford University
    Press.
Heil, John. 2005. 'Dispositions.' *Synthese* 144: 343–56.
Heil, John. 2012. 'Are Four Categories Two too Many?' In Tahko (ed.), 2012.
Heil, J. and Mele, A. (eds) 1993. *Mental Causation*. Oxford: Clarendon Press.
Hempel, Carl. 1965. *Aspects of Scientific Explanation*. New York: Free Press.
Hintikka, J. (ed.) 1975. *Rudolf Carnap, Logical Empiricist*. Dordrecht: D. Reidel.
Hirsch, E. 1993. *Dividing Reality*. Oxford: Oxford University Press.
Hirsch, E. 2002. 'Quantifier Variance and Realism.' *Philosophical Issues* 12:
    52–73.
Hirsch, E. 2009. 'Ontology and Alternative Languages.' In Chalmers et al. (eds),
    2009: 231–59.
Hochberg, Herbert. 2001. 'A Refutation of Moderate Nominalism.' In his *Russell,
    Moore and Wittgenstein: The Revival of Realism*. Frankfurt am Maine:
    Hänsel-Hohenhausen.
Hochberg, Herbert. 2004. 'Relations, Properties and Predicates.' In Hochberg
    and Mulligan (eds), 2004: 17–53.

Hochberg, H. and Mulligan, K. (eds) 2004. *Relations and Predicates.* Heusenstamm: Ontos Verlag.

Hoeltje, M., Schnieder, B. and Steinberg, A. (eds) 2013. *Varieties of Dependence: Ontological Dependence, Grounding, Supervenience, Response-Dependence.* Munich: Philosophia Verlag.

Hofweber, T. 2009. 'Ambitious, yet Modest, Metaphysics.' In Chalmers et al. (eds), 2009: 260–89.

Honderich, Ted. 1988. *Mind and Brain. A Theory of Determinism.* Volume 1. Oxford: Clarendon Press.

Hudson, H. 2007. 'Simples and Gunk.' *Philosophy Compass* 2: 291–302.

Hume, David. 1777. (3rd edn 1975.) *An Enquiry Concerning Human Understanding.* Oxford: Clarendon Press.

Husserl, E. 1970 [1900]. *Logical Investigations.* Translated by J. N. Findlay. London: Routledge and Kegan Paul.

Hüttemann, Andreas. 2013. 'A Disposition-based Process-theory of Causation.' In Mumford and Tugby (eds), 2013: 101–22.

Jespersen, B. and Duži, M. (eds) 2015. Hyperintensionality. Special Issue of *Synthese.*

Keinänen, M. and Hakkarainen, J. 2014. 'The Problem of Trope Individuation: A Reply to Lowe.' *Erkenntnis* 79: 65–79.

Kim, Jaegwon. 1982. 'Psychophysical Supervenience.' *Philosophical Studies* 41: 51–70. Reprinted in Kim, 1993: 175–93.

Kim, Jaegwon. 1993a. *Supervenience and Mind.* Cambridge: Cambridge University Press.

Kim, Jaegwon. 1993b. 'The Nonreductivist's Troubles with Mental causation.' In Heil and Mele (eds), 1993: 189–210.

Kim, Jaegwon. 1998. *Mind in a Physical World.* Cambridge, MA: MIT Press.

Kripke, Saul. 1980. *Naming and Necessity.* Cambridge, MA: Harvard University Press.

Kuhlmann, Meinard. 1999. 'Quanta and Tropes: Trope Ontology as Descriptive Metaphysics of Quantum Field Theory.' In Meixner and Simons (eds), 1999: 338–43.

Kuhlmann, Meinard. 2010. *The Ultimate Constituents of the Material World – In Search of an Ontology for Fundamental Physics.* Frankfurt: ontos-verlag.

Küng, Guido. 1967. *Ontology and the Logistic Analysis of Language.* Dordrecht: D. Reidel.

Ladyman, James and Ross, D., with J. Collier and D. Spurrett. 2007. *Every Thing Must Go.* Oxford: Oxford University Press.

Langton, Rae and Lewis, D. 1998. 'Defining "Intrinsic".' *Philosophy and Phenomenological Research* 58: 333–45.

Langton, Rae and Lewis, D. 2001. 'Marshall and Parsons on "Intrinsic".' *Philosophy and Phenomenological Research* 63: 353–5.

Lewis, David. 1969. 'Policing the *Aufbau.*' *Philosophical Studies* 20: 13–17.

Lewis, David. 1973. *Counterfactuals.* Oxford: Blackwell.

Lewis, David. 1979. 'Counterfactual Dependence and Time's Arrow.' *Noûs* 13: 455–76.

Lewis, David. 1983a. 'New Work for a Theory of Universals.' *Australasian Journal of Philosophy* 61: 343–77. Reprinted in Mellor and Oliver (eds), 1997: 190–227.

Lewis, David. 1983b. 'Extrinsic Properties.' *Philosophical Studies* 44: 197–200.

Lewis, David. 1986. *On the Plurality of Worlds*. Oxford: Blackwell.

Lewis, David. 2002. 'Tensing the Copula.' *Mind* 111: 1–14.

Lewis, David. 2009. 'Ramseyan Humility.' In Braddon-Mitchell and Nolan (eds), 2009: 203–22.

Loux, Michael J. (ed.) 2001. *Metaphysics: Contemporary Readings*. London: Routledge.

Loux, M. and Zimmerman, D. (eds) 2003. *The Oxford Handbook of Metaphysics*. Oxford: Oxford University Press.

Lowe, E. J. 1989. 'What is a Criterion of Identity?' *Philosophical Quarterly* 39: 1–21.

Lowe, E. J. 1998. *The Possibility of Metaphysics*. Oxford: Clarendon Press.

Lowe, E. J. 2002a. *A Survey of Metaphysics*. Oxford: Oxford University Press.

Lowe, E. J. 2002b. 'Kinds, Essence and Natural Necessity.' In Bottani et al. (eds), 2002: 189–206.

Lowe, E. J. 2003. 'Individuation.' In Loux and Zimmerman (eds), 2003: 75–95.

Lowe, E. J. 2006. *The Four-Category Ontology*. Oxford: Oxford University Press.

MacDonald, Cynthia. 1989. *Mind-Body Identity Theories*. London: Routledge.

Manley, D. 2002. 'Properties and Resemblance Classes.' *Noûs* 36: 75–96.

Marcel, A. J. and Bisiach, E. (eds) 1988. *Consciousness in Contemporary Science*. Oxford: Oxford University Press.

Marmadoro, Anna. 2010a. 'Do Powers need Powers to make them Powerful?' In Marmadoro (ed.), 2010b: 337–52.

Marmadoro, Anna. (ed.) 2010b. *The Metaphysics of Powers: their grounding and their manifestation*. London: Routledge.

Marshall, Dan. 2009. 'Can '*Intrinsic*' be Defined Using only Broadly Logical Notions?' *Philosophy and Phenomenological Research* 78: 647–72.

Marshall, Dan. 2012. 'Analyses of Intrinsicality in Terms of Naturalness.' *Philosophy Compass* 7/8: 531–42.

Marshall, Dan and Parsons, J. 2001. 'Langton and Lewis on "Intrinsic".' *Philosophy and Phenomenological Research* 63: 347–51.

Martin, C. B. 1980. 'Substance Substantiated.' *The Australasian Journal of Philosophy* 58: 3–10.

Martin, C. B. 1997. 'On the Need for Properties: The road to Pythagoreanism and back.' *Synthese* 112: 193–231.

Martin, C. B. 2008. *The Mind in Nature*. Oxford, Oxford University Press.

Martin, C. B. and Heil, J. 1999. 'The Ontological Turn.' *Midwest Studies in Philosophy* 23: 34–60.

Maurin, Anna-Sofia. 2002. *If Tropes*. Dordrecht, The Netherlands: Kluwer Academic Publishers.

Maurin, Anna-Sofia. 2005. 'Same but Different.' *Metaphysica* 6: 129–46.

Maurin, Anna-Sofia. 2010a. 'A World of Tropes?' In Vanderbeeken and D'Hooghe (eds), 2010: 107–30.

Maurin, Anna-Sofia. 2010b. 'Trope Theory and the Bradley Regress.' *Synthese* 175: 311–26.

Maurin, Anna-Sofia. 2011. 'An Argument for the Existence of Tropes.' *Erkenntnis* 74: 69–79.

McDaniel, K. 2007. 'Extended Simples.' *Philosophical Studies* 133: 131–41.

McGowan, M.-K. 2002. 'The Neglected Controversy over Metaphysical Realism.' *Philosophy* 77: 5–21.

McKitrick, Jennifer. 2013. 'How to Activate a Power.' In Mumford and Tugby (eds) 2013: 123–37.

Meinong, A.1904. 'Über Gegenstandstheorie.' In *Untersuchungen zur Gegenstandstheorie und Psychologie*. Leipzig: Barth. Translated as 'On the Theory of Objects' in Chisholm (ed.) 1960: 76–117.

Meixner, U. and Simons, P. M. (eds) 1999. *Contributions of the Austrian Ludwig Wittgenstein Society* (22nd International Wittgenstein Symposium).

Mellor, D. H. 1993. 'Properties and Predicates.' In Bacon, Campbell and Reinhardt (eds), 1993: 101–12.

Mellor, D. H. and Oliver, A. (eds) 1997. *Properties*. Oxford: Oxford University Press.

Miščević, Nenad. 1997. 'Why Pettit Cannot have it Both Ways.' *Language, Mind and Society*. Maribor, Slovenia: University of Maribor Press: 115–22.

Molnar, G. 2003. *Powers: A Study in Metaphysics*. Oxford: Oxford University Press.

Moore, G. E. 1959. 'A Defence of Common Sense.' In his *Philosophical Papers*. London: Allen and Unwin: 32–59.

Morganti, Matteo. 2007. 'Resembling Particulars: What Nominalism?' *Metaphysica* 8: 165–78.

Morrison, M. 1994. 'Unified Theories and Disparate Things.' *Proceedings of the Biennial Meeting of the Philosophy of Science Association* 2: 365–73.

Mulligan, K., Simons, P. and Smith, B. 1984. 'Truth-Makers.' *Philosophy and Phenomenological Research* 44: 287–321.

Mumford, S. 1998. *Dispositions*. Oxford: Oxford University Press.

Mumford, S. 2004. *Laws in Nature*. London: Routledge.

Mumford, S. 2009. 'Passing Powers Around.' *The Monist* 92: 94–111.

Mumford, S. and Anjum, R. L. 2010. 'A Powerful Theory of Causation.' In Marmadoro (ed.), 2010b.

Mumford, S. and Anjum, R. L. 2011. 'Dispositional Modality.' In C. F. Gethmann (ed.), *Lebenswelt und Wissenschaft, Deutsches Jahrbuch Philosophy 2*. Meiner Verlag: 380–94.

Mumford, S. and Tugby, M. (eds) 2013. *Metaphysics and Science*. Oxford: Oxford University Press.

Nagel, Ernest. 1961. *The Structure of Science*. London: Routledge and Kegan Paul.

Newton-Smith, W. H. 1980. *The Structure of Time*. London: Routledge.

Nicolis, G. and Prigogine, I. 1977. *Self-Organization in Non-Equilibrium Systems*. New York: Wiley.

Nolan, Daniel. 2014. 'Hyperintensional Metaphysics.' *Philosophical Studies* 171: 149–60.

Orilia, Francesco. 2006. 'States of Affairs. Bradley vs. Meinong.' In Raspa (ed.), 2006: 213–38.

Paseau, Alexander. 2012. 'Resemblance Theories of Properties.' *Philosophical Studies* 157: 361–82.

Pautz, Adam. 1997. 'An Argument Against Armstrong's Analysis of the Resemblance of Universals.' *The Australasian Journal of Philosophy* 75: 109–11.

Pettit, P. 1998. 'Noumenalism and Response-dependence.' *The Monist* 81: 112–32.

Plato. *The Republic.*

Plato. *Parmenides.*

Price, H. H. 1953. 'Universals and Resemblance.' In his *Thinking and Experience.* Cambridge, MA: Harvard University Press: Ch. 1. Reprinted in Loux (ed.), 2001: 20–41.

Price, Huw. 2009. 'Metaphysics after Carnap: The Ghost who Walks?' In Chalmers et al. (eds), 2009: 320–46.

Priest, Graham. 1997. 'Sylvan's Box: A Short Story and Ten Morals.' *Notre Dame Journal of Formal Logic* 38: 573–81.

Prigogine, I. and Stengers, I. 1984. *Order Out of Chaos.* New York: Bantam Books.

Proust, J. 1989. *Questions of Form: Logic and the Analytic Proposition from Kant to Carnap.* Translated by Anastasios Albert Brenner. Minneapolis: University of Minnesota Press.

Psillos, S. 2006. 'What do Powers do when they are not Manifested?' *Philosophy and Phenomenological Research* 72: 137–56.

Putnam, Hilary. 1975a. 'Philosophy and our Mental Life.' In *Mind, Language and Reality: Philosophical Papers*, Vol. 2. Cambridge: Cambridge University Press.

Putnam, Hilary. 1975b. 'The Meaning of "Meaning".' *Minnesota Studies in the Philosophy of Science* 7: 131–93.

Putnam, Hilary. 1975c. *Mathematics, Matter and Method.* Cambridge: Cambridge University Press.

Putnam, Hilary. 1979. 'Philosophy of Logic.' Reprinted in *Mathematics Matter and Method: Philosophical Papers, Vol. 1*, 2nd edn. Cambridge: Cambridge University Press: 323–57.

Putnam, Hilary. 1981. *Reason, Truth and History.* Cambridge: Cambridge University Press.

Quine, W. V. 1948. 'On What There Is.' *The Review of Metaphysics.* Reprinted in Quine, 1953: 1–19.

Quine, W. V. 1951. 'Two Dogmas of Empiricism.' *The Philosophical Review.* Reprinted in Quine, 1953: 20–46.

Quine, W. V. 1953 (2nd edn 1960). *From a Logical Point of View.* Cambridge, MA: Harvard University Press.

Quine, W. V. 1960. *Word and Object.* Cambridge, MA: MIT Press.

Quine, W. V. 1969. 'Ontological Relativity.' In *Ontological Relativity and Other Essays.* New York: Columbia University Press: 26–68.

Quine, W. V. 1975. 'On the Individuation of Attributes.' Reprinted in Quine, 1981: 100–12.

Quine, W. V. 1976. 'Carnap and Logical Truth.' In his *Ways of Paradox and Other Essays.* Cambridge, MA: Harvard University Press: 107–32.

Quine, W. V. 1981. *Theories and Things.* Cambridge, MA: Harvard University Press.

Quine, W. V. 1990a. *Pursuit of Truth.* Cambridge, MA: Harvard University Press.

Quine, W. V. 1990b. 'Three Indeterminacies.' In Barrett and Gibson (eds), 1990: 1–16.

Ramsey, F. 1925. 'Universals.' *Mind* 34: 401–17.

Ramsey, F. 1978. *Foundations: Essays in Philosophy, Logic, Mathematics and Economic.* London: Routledge and Kegan Paul.

Raspa, V. (ed.) 2006. *Meinongian Issues in Contemporary Italian Philosophy.* Frankfurt: Ontos.

Rodriguez-Pereyra, G. 2002. *Resemblance Nominalism.* Oxford: Oxford University Press.

Rosen, G. 2010. 'Metaphysical Dependence: Grounding and Reduction.' In Hale and Hoffmann (eds), 2010: 109–36.

Russell, Bertrand. 1905. 'On Denoting.' *Mind* 14: 479–93

Russell, Bertrand. 1912a. *The Problems of Philosophy.* Oxford: Oxford University Press.

Russell, Bertrand. 1912b. 'On the Notion of Cause.' *Proceedings of the Aristotelian Society* 13: 1–26.

Sankey, H. (ed.) 1999. *Causation and Laws of Nature.* Dordrecht: Kluwer.

Schaffer, Jonathan. 2004. 'Two Conceptions of Sparse Properties.' *Pacific Philosophical Quarterly* 85: 92–102.

Schaffer, Jonathan. 2005. 'Quidditistic Knowledge.' *Philosophical Studies* 123: 1–32.

Schaffer, Jonathan. 2009. 'On What Grounds What.' In Chalmers et al. (eds), 2009: 347–83.

Schaffer, Jonathan. 2010. 'Monism: The Priority of the Whole.' *Philosophical Review* 119: 31–76.

Schmidt, Martin. 2005. 'Can Bundle Theory Explain Individuation?' *Organon F*: 62–71.

Schrenk, M. 2010. 'The Powerlessness of Necessity.' *Noûs* 44: 725–39

Schroer, Robert. 2013. 'Can a Single Property be both Dispositional and Categorical? The "Partial Consideration Strategy" Partially Considered.' *Metaphysica* 14: 63–77.

Shoemaker, S. 1969. 'Time without Change.' *Journal of Philosophy* 66: 363–81.

Shoemaker, S. 1980. 'Causality and Properties.' Reprinted in Mellor and Oliver (eds), 1997: 228–54.

Sider, Theodore. 1993. 'Intrinsic Properties.' *Philosophical Studies* 83: 1–27.

Sider, Theodore. 1995. 'Sparseness, Immanence and Naturalness.' *Noûs* 29: 360–77.

Sider, Theodore. 1996. 'Naturalness and Arbitrariness.' *Philosophical Studies* 81: 283–301.

Sider, Theodore. 2001. 'Maximality and Intrinsic Properties.' *Philosophy and Phenomenological Research* 63: 357–64.

Simons, P. 1994. 'Particulars in Particular Clothing.' *Philosophy and Phenomenological Research* 54: 553–75.

Sosa, E. and Tooley, M. (eds) 1993. *Causation.* Oxford: Oxford University Press.

Stout, G. F. 1921–3. 'The Nature of Universals and Propositions.' *Proceedings of the British Academy* 10: 157–72.

Stout, G. F. 1923. 'Are the Characteristics of Particular things Universal or Particular?' *Proceedings of the Aristotelian Society* (Supp. Vol.) 3: 114–22.

Strawson, G. 2008. 'The Identity of the Categorical and the Dispositional.' *Analysis* 68: 271–82.

Swoyer, Chris. 1982. 'The Nature of Natural Laws.' *Australasian Journal of Philosophy* 60: 203–23.

Swoyer, Chris. 1996. 'Theories of Properties: From Plenitude to Paucity.' *Philosophical Perspectives* 10: 243–64.

Tahko, Tuomas E. (ed.) 2012. *Contemporary Aristotelian Metaphysics.* Cambridge: Cambridge University Press.

Taylor, Barry. 1993. 'On Natural Properties in Metaphysics.' *Mind* 102: 81–100.

Tooley, M. 1977. 'The nature of laws.' *Canadian Journal of Philosophy* 7: 667–98.

Trogdon, Kelly. 2009. 'Monism and Intrinsicality.' *Australian Journal of Philosophy* 87: 127–48.

Trogdon, Kelly. 2013. 'An Introduction to Grounding.' In Hoeltje et al. (eds), 2013: 97–122.

Vallentyne, Peter. 1997. 'Intrinsic Properties Defined.' *Philosophical Studies* 88: 209–19.

Vallicella, William F. 2002. 'Relations, Monism, and the Vindication of Bradley's Regress.' *Dialectica* 56: 3–35.

Van Fraassen, Bas C. 1980. *The Scientific Image.* Oxford: Oxford University Press.

Vanderbeeken, Robrecht and D'Hooghe, Bart. (eds) 2010. *Worldviews, Science, and Us: Studies of Analytic Metaphysics.* Brussels: World Scientific Publishers, 2010.

Vetter, B. 2011. 'Recent Work: Modality without Possible Worlds.' *Analysis Reviews* 71: 742–54.

Vetter, B. 2015. *Potentiality: From Dispositions to Modality.* Oxford: Oxford University Press.

Weatherson, B. 2001. 'Intrinsic Properties and Combinatorial Principles.' *Philosophy and Phenomenological Research* 63: 365–80.

Weatherson, Brian and Marshall, Dan. 2014. 'Intrinsic vs. Extrinsic Properties.' *The Stanford Encyclopedia of Philosophy* (Fall 2014 Edition). Edward N. Zalta (ed.), http://plato.stanford.edu/archives/fall2014/entries/intrinsic-extrinsic/

Williams, Donald C. 1953. 'On the Elements of Being: I.' Reprinted in Mellor and Oliver (eds), 1997: 112–24.

Williams, Donald C. 1963. 'Necessary Facts.' *The Review of Metaphysics* 16: 601–25.

Wilson, Alastair. 2013. 'Schaffer on Laws of Nature.' *Philosophical Studies* 164: 653–67.

Witmer, D. G., Butchard, W. and Trogdon, K. 2005. 'Intrinsicality without Naturalness.' *Philosophy and Phenomenological Research* 70: 326–50.

Yablo, S. 1999. 'Intrinsicness.' *Philosophical Topics* 26: 479–505.

Yagisawa, T. 1988. 'Beyond Possible Worlds.' *Philosophical Studies* 53: 175–204.

Zalta, Edward N. 1983. *Abstract Objects: An Introduction to Axiomatic Metaphysics.* Dordrecht: D. Reidel.

Zalta, Edward N. 1988. *Intensional Logic and the Metaphysics of Intentionality.* Cambridge, MA: MIT Press.

Index

actualism (modal) 3, 21–4, 96, 144, 151–3, 181, 183–8
alien properties (uninstantiated properties) 21–4, 54, 147, 151–6, 180–1, 183, 187, 200
Armstrong, D. 8, 12, 14, 17, 19–21, 23–7, 32–3, 42, 44, 50, 55, 71, 87, 103–5, 142, 148, 160, 169, 175–6, 178, 211–12

Carnap, R. 71–2, 78–81, 196–7
categorical properties 139–50, 154, 158, 173, 178, 182–3 *see also* quiddity
Cause–Law Thesis 171, 176–81, 210–11
coextension problem 72–7
combinatorialism 148–9, 157, 182–3
companionship problem 78–80, 83–5
contingentism 141, 145–59, 173

D-relational properties 129–33, 135, 137
Davidson, D. 170, 203, 211–12
determinable/determinate properties 26–7, 79, 101

Ehring, D. 44, 49–50, 56–60, 63
Elgin, Catherine Z. 199, 208
epiphenomenal properties 110
extrinsicality *see* D–relational properties

Frege, G. 32–3, 98

Goodman, Nelson 77–82
grounding and Intrinsicality 124–8, 132

haecceity 146, 154
Hirsch, Eli 123, 196, 203, 205
hyperintensionality 100–1, 125, 133

ideal standards 20–1
identity and individuation criteria (properties) 98–101, 108–11
imperfect community, the problem of 78, 80–5
individuation (tropes) 51–60
instantiation regress 28–33
intrinsic properties 113–37

Kim, J. 118–19

laws of nature 13–14, 122, 150, 156, 170–1, 176–81, 202–3, 210–12
and singular causation 171–6
Lewis, David 23, 54, 74–5, 84, 87, 102–5, 108, 146, 148, 150, 158, 178, 183, 201, 207, 211
on intrinsic properties 118–24, 126–8, 132–3, 137
loneliness 118–28
Lowe, E. J. 9, 19–20, 22

Maurin, Anna-Sofia 43–7, 51, 60–2
Mumford, S. 115, 179, 181

nominalism 68–9

Orilia, F. 29–30

pandispositionalism 159–63
piling (tropes) 57–8
Price, H. H. 14, 26, 72
principle of instantiation 3, 21–3, 30, 183, 200
Putnam, Hilary 96, 105, 117, 199, 207

quiddity 141, 146–8, 154, 158, 163, 200–2
Quine, W. V. 18, 74, 96, 98–9, 104, 132, 195–7, 199

relations 1–2
resemblance (inexact) and universals 24–7
resemblance (tropes) 42–51
resemblance regress 12–13, 47, 50
Rodriguez–Pereyra, G. 76, 82–5
Russell, B. 12–18, 34

scattering (tropes) 58–9
Schaffer, J. 53–9, 103–4, 108, 144, 149, 156, 201
schmarge 152–6 *see also* alien properties
self–instantiation 33–4
singular causation 62–3, 169–76, 210
space–time, the nature of 14, 15–17, 20, 53, 115

sparse properties 102–8, 156, 160, 193–4, 197–208
strange science 203–7
supervenience 117, 136–7, 167, 176
swapping (tropes) 55–7

Taylor, Barry 123, 199, 208, 211
truth-maker theory and tropes 32, 44–7, 61
truth-makers and counterfactuals 144–5, 150, 154, 157–8, 178, 181–7
truth-makers and singular causation 172

Vetter, B. 185–8